COMPUTER-AIDED CIRCUIT ANALYSIS USING PSpice®

SECOND EDITION

WALTER BANZHAF

Ward College of Technology
University of Hartford
West Hartford, Connecticut

REGENTS/PRENTICE HALL , Englewood Cliffs, N.J. 07632

Library of Congress Cataloging-In-Publication Data

Banzhaf, Walter.
    Computer-aided circuit analysis using PSpice / Walter Banzhaf. — 2nd. ed.
      p.   cm.
    Rev. ed. of: Computer-aided circuit analysis using SPICE. c1989.
    Includes bibliographical references and index.
    ISBN 0-13-159534-2
    1. Electrical circuit analysis—Data processing.   2. Pspice
(Computer program) I. Banzhaf, Walter. Computer-aided circuit
analysis using SPICE. II. Title.
    TK454.B33   1992
621.381'5'02855369—dc20   91-44940
                                              CIP

Editorial/production supervision: *Mary Carnis*
Cover design: *Wanda Lubelska*
Manufacturing buyer: *Ed O'Dougherty*
Prepress buyer: *Ilene Levy*
Page Layout: *Marty Behan*

IBM® is a registered trademark of International Business
Machines Corporation. PSpice® is a registered trademark of
MicroSim Corporation.

The representation of screens from PSpice and its options used by permission of MicroSim Corporation.
Copyright 1990 MicroSim Corporation. This material is proprietary to MicroSim
Corporation. Unauthorized use, copying, or duplication is strictly prohibited. PSpice is a registered trade-
mark of MicroSim Corporation. StmEd, Stimulus Editor, Probe, Parts, Monte Carlo, Analog Behavioral
Modeling, Device Equations, Digital Similation, Digital Files, and Filter Designer are trademarks of
MicroSim Corporation.

Printed in the United States of America

10 9 8 7 6 5 4 3 2 1

ISBN 0-13-159534-2

Prentice-Hall International (UK) Limited, *London*
Prentice-Hall of Australia Pty, Limited, *Sydney*
Prentice-Hall Canada Inc., *Toronto*
Prentice-Hall Hispanoamericana, S.A., *Mexico*
Prentice-Hall of India Private Limited, *New Delhi*
Prentice-Hall of Japan, Inc., *Tokyo*
Simon & Schuster Asia Pte. Ltd., *Singapore*
Editora Prentice Hall do Brasil, Ltda., *Rio de Janeiro*

CONTENTS

Contents

PREFACE

The first edition of this book was written with the express purpose of making it possible (and easy) for anyone in the electronics field to start using SPICE to analyze electric or electronic circuits in a very short time. Engineering technology students at my college originally used SPICE on a mainframe computer and now use PSpice. They were able to master use of PSpice by following examples which they received in class. Each example contained a schematic diagram of the circuit to be analyzed, the input file which described the circuit to PSpice, and the output file or graphics created by PSPice with a discussion of its contents. After making a few of the classical errors (e.g. typing a letter "O" when the digit "0" (zero) was intended), most students were quite comfortable doing PSpice analyses completely on their own.

This second edition benefits from the many helpful suggestions I received. I'm grateful to students and faculty at my college, colleagues at other institutions, and reviewers who were kind enough to share their perceptions with me. Foremost, PSpice (which is derived from SPICE and runs on a PC) is used exclusively as the subject of this book. MicroSim Corporation has been refreshingly supportive of higher education by making available to faculty an evaluation version of PSpice (with its many options) for classroom use. Faculty are welcome and encouraged to share the evaluation software with students. As a result, PSpice is widely used in colleges nationwide.

The PSpice graphics post-processor Probe has been used throughout this second edition to make high-quality graphs of analysis results. All chapters in the book have been revised, and many additions were made. Switches, macro models of operational amplifiers, and the PSpice options Analog Behavioral Modeling and Parts are now included. A quick reference table of circuit element lines and control lines is included as an appendix.

This book is designed to be of immediate use to two types of readers. The first comprises engineering and engineering technology students, for whom this book could be a supplemental text for any circuits or electronics course. In addition to learning classical methods of circuit analysis and design, students also need to learn how to perform circuit analysis on computers in order to be prepared for industry. An added benefit of simulating circuit behavior on a computer is that students can gain great insight into circuit behavior which would be otherwise unattainable due to the sheer tedium of the mathematical operations necessary. For example, a student having spent the better part of an hour analyzing a transistor amplifier with a BJT whose beta is 80 is not likely to repeat the exercise with the beta changed to 40 just to satisfy intellectual curiosity. With PSpice, repeating the analysis would take only a few seconds. Also, by using computer simulations to check the work they do by hand, students rapidly gain confidence in their own analytical abilities.

Practicing engineers, technologists and technicians who are already competent at circuit analysis and design and want to learn how to use PSpice will also find this book useful. It will provide details about PSpice which will help them to use the computer to confirm results obtained by hand, to save time in the laboratory, and to go from a concept to a working prototype quickly. The wide range of example circuits will allow circuit designers to learn the PSpice syntax quickly and begin using a PC with a minimum of delay.

This second edition has a large number of examples which illustrate nearly all the capabilities of PSpice. A first-year student may be able to use only the DC and AC analysis examples initially. As that student's knowledge of electronics advances, more and more of the book will be appropriate and useful. Conversely, a practicing engineer, technologist or technician may wish to jump to the more advanced topics right away. It is advisable for that reader at least to skim through the preliminary chapters to see both the pitfalls (not many, but worth knowing about) and the extensive capabilities of PSpice.

The part of the book that ties it all together, Chapter 12, presents 30 examples of PSpice analysis. Each example contains a schematic diagram of a circuit to be analyzed, an input file submitted to PSpice for analysis, the results in a graph made with Probe (or an output file created by PSpice) and a discussion of key points illustrated by the example.

Each example in the text can be considered to be a practice problem, and it is expected that readers will have no shortage of circuits to analyze from courses being taken (if students) or from daily work (if professionals). Only by applying PSpice to your own circuits will you fully understand its capabilities and achieve a mastery of it for your purposes.

I am grateful to all those at the University of Hartford who supported this effort and helped make it happen. This certainly includes my students from whom I learn so much about learning. Mary Carnis of Prentice Hall has been a delight to work with on production of both editions of this book. My wife, Mattie (an eagle-eyed proofreader), and my children Amy and Jeremy have been supportive and understanding always; without their help I could not have written this book. Thank you all.

I'd truly appreciate receiving any suggestions, corrections or predilections about this book, at the address below.

Walter Banzhaf
Ward College of Technology
University of Hartford
West Hartford, Connecticut 06117

Chapter 1

INTRODUCTION TO COMPUTER-AIDED CIRCUIT ANALYSIS AND PSpice

1.1 ACKNOWLEDGMENTS

SPICE was developed by the CAD (Computer-Aided Design) Group of the University of California, Berkeley, and is the sole property of the Regents of the University of California. The author is grateful to the University of California for granting permission to use excerpts from its Electronics Research Laboratory memoranda in this book.

PSpice was developed by MicroSim Corporation of Irvine, California. The representation of screens from PSpice and its options are used by permission of MicroSim Corporation. The author, and indeed thousands of students and faculty studying electronics, are indebted to MicroSim for its enlightened policy of making a powerful evaluation version of PSpice software available for educational use.

1.2 THE NEED FOR COMPUTER-AIDED CIRCUIT ANALYSIS

Circuit analysis is a necessary part of circuit design. Once a design for a circuit has been determined, the soundness of the design must be tested to ensure that the circuit does indeed perform as required. Often this involves testing for DC operating point and performance, with signal applied, over a range of DC supply voltages, input signal levels, and temperatures. The time-honored way to do this was to build a prototype of the circuit, send it off to the laboratory, and invest large amounts of time and money in putting the circuit through its paces in the hope that it would perform as desired.

SPICE was developed in the 1970s to allow circuits to be simulated on a mainframe computer, thus saving huge amounts of time and money in the early development stages of circuit design.

It is not necessary to understand the algorithms and mathematical techniques that are utilized by SPICE to analyze circuits in order to be successful at solving circuit problems with SPICE, any more than one must have a working knowledge of internal combustion engines to drive a car. However, it is vitally important for anyone using SPICE (or any other computer analysis/design program) to be a competent practitioner in the field. Only in that way can one check the correctness and reasonableness of the computer's answers by estimation using sound engineering judgement and/or by working a sample problem through by conventional means.

1.3   A LITTLE BIT ABOUT SPICE & PSpice

This book concentrates on PSpice, a PC version of SPICE which runs on personal computers and offers many improvements over SPICE. SPICE is an acronym for Simulation Program with Integrated Circuit Emphasis, which along with its derivatives is the most popular computer program in the world today for predicting the behavior of electronic circuits. It was developed by the Integrated Circuits Group of the Electronics Research Laboratory and the Department of Electrical Engineering and Computer Sciences at the University of California, Berkeley, California. The person credited with originally developing SPICE is Dr. Lawrence Nagel, whose PhD thesis describes the algorithms and numerical methods used in SPICE. SPICE has undergone many changes since it was first developed, and it continues to evolve.

SPICE is a large (over 17,000 lines of FORTRAN source code), powerful and extremely versatile industry-standard program for circuit analysis and IC design. A significant number of companies have customized Berkeley SPICE for in-house circuit development work. Many software packages based on SPICE have been developed which use the SPICE2 program from Berkeley as the core for performing circuit analysis. Most of these include useful programs added to make the complete package easier to use. For example, SPICE2 is not interactive, it does not have the capability to reference a library of semiconductor components, and its graphs are made on a line printer using ASCII (American Standard Code for Information Interchange) symbols. Some of the commercially available packages are interactive, do include extensive libraries of parts, and have graphics post-processors which make professional-looking graphs.

SPICE and other packages based on it will be around and widely used for many years to come. As you read this book, you will see that SPICE, though developed for the design of integrated circuits, can be used to solve a great variety of non-IC circuit problems involving power supplies, three-phase power systems, transmission lines and non-linear components, to name a few. Anyone performing circuit analysis or design should have a working knowledge of SPICE in order to save time and money and to gain insight into circuit behavior by answering "what if" questions with a computer simulation. "What if" questions are frequently not answered if the tedium involved in doing so outweighs the curiosity of the person asking the question.

Students will find PSpice to be an important tool for learning circuit analysis and design, and for testing electronic circuits in ways they could not easily do in most college laboratories. By learning a version of SPICE, students will also be preparing for the kind of circuit simulation they will encounter in industy. Faculty must be careful, however, to ensure that students are competent in traditional methods of circuit analysis before exposing them to computer methods.

1.4 INSTALLING PSpice ON YOUR COMPUTER

This section tells, in general terms, how to install the evaluation version of PSpice on a PC using DOS. Updated versions are released by MicroSim Corporation on a periodic basis; refer to the instructions that come with the software and the PSpice User's Guide, which may be purchased from MicroSim Corporation, for the latest details.

1.4.1   What Your Computer Must Have

In order to run the evaluation version of PSpice (including Probe and Parts), your computer must have at least the following:

> 640 Kilobytes of memory
> A hard (fixed) disk
> MS-DOS 3.0 (or higher version) operating system

Desirable, but not necessary, are a color monitor, graphics capability for the PC, a printer that will do graphics, and a math coprocessor (which will speed things up considerably).

1.4.2   Putting PSpice Onto The Hard Disk

This part is really very straightforward. If you are unfamiliar with any of the operations below, either refer to the DOS manual for your version of DOS or ask someone more knowledgable for assistance.

1) Make a subdirectory on your hard disk (a good name to use is PSPICE):
   If the hard disk drive letter is called C, after your PC has booted either C> or C:\> will appear on the monitor. To make the subdirectory, key in MD PSPICE and hit ENTER.

2) Copy all the files on the floppy disks containing the evaluation version to the PSpice subdirectory:
   Put the first PSpice floppy disk in drive A, and key in COPY A:*.*   C:\PSPICE and hit ENTER. Repeat this step with each of the PSpice floppy disks.

Note: If your floppy drive is a 5.25-inch, 360 Kbyte size, and the PSpice floppy disks are 5.25-inch, 1.2 Mbyte size, then one solution is to:
   a) find a PC which has a 1.2 Mbyte 5.25-inch drive
   b) put PSpice onto that PC's hard drive by following steps 1 and 2 above
   c) use BACKUP from your DOS version to backup all PSpice files from that PC's PSpice subdirectory onto 360 Kbyte floppy disks
   d) use RESTORE from your DOS version to restore all backed

up PSpice files from the 360 Kbyte floppy disks onto your PC's PSpice subdirectory

1.4.3   One Last Thing You <u>Must</u> Do - CONFIG.SYS

PSpice needs to open a goodly number of temporary (.TMP) files to store data while it is analyzing a circuit. These files are deleted automatically when PSpice completes the analysis (you may find some of these .TMP files in your PSpice subdirectory if power goes out in the middle of an analysis). PSpice also needs a minimum number of disk buffers to be allocated in your PC's memory for the same purpose.

At the time your PC is turned on, DOS will look for a file called CONFIG.SYS in the root directory. If the PC finds it, the PC will (generally) do what CONFIG.SYS says; if not, DOS will decide what to do. What DOS decides to do by default is often not right for PSpice. So it is up to you to make sure that:

   1) a file exists in the root directory called CONFIG.SYS
   2) CONFIG.SYS contains at least the following two lines:
         BUFFERS = 20
         FILES = 20

CONFIG.SYS may already exist, and may contain quite a few lines. If so, use a text editor (not a word processor) to add or change lines so that BUFFERS = 20 and FILES = 20 appear (more than 20 is OK but not necessary for PSpice). If CONFIG.SYS does not exist, create it and put these two lines in it. Since CONFIG.SYS is read by DOS only once, at boot-up, <u>after you have changed CONFIG.SYS you must re-boot the PC</u>.

CHAPTER SUMMARY

SPICE was developed for analyzing IC designs by the CAD Group of UC/Berkeley in the 1970s.

SPICE (and derivatives of it) are widely used in industry and universities. Though developed for IC design, it can be a useful tool for both the student and practicing circuit designer with many non-IC circuits.

SPICE can quickly give answers to "what-if" kinds of questions that would be all but unanswerable without computer-aided circuit analysis.

Those who use SPICE must be competent in circuit analysis to be able to judge the reasonableness of computer-generated results.

Installing the evaluation version of PSpice on the hard disk of a PC is a straightforward task. The CONFIG.SYS file in the root directory must be customized for PSpice.

Chapter 2

GROUND RULES OF PSpice
AND BASIC ELEMENT
LINE RULES

2.1 INTRODUCTION

In order to perform circuit analysis, one must first have a circuit to analyze. The purpose of this profound statement is to reinforce the idea that the circuit design must be done by a person. PSpice can only simulate how a given circuit will behave; it can't suggest ways to improve the performance of that circuit, and it certainly can't synthesize a circuit when given the design criteria.

SPICE is not an interactive program, while PSpice may be run in an interactive mode. This book will describe how to perform circuit analysis in the non-interactive mode. The circuit and the types of analysis to be done are described to PSpice in an input file. This input file is created using a text editor on a computer, and is submitted to PSpice for analysis. The results of the analysis (or helpful error messages, if errors existed in the input file) appear in an output file written by PSpice. If the results indicate that the circuit needs to be changed, the circuit designer modifies the circuit, edits the input file, runs PSpice again, and examines the output file. This process can be repeated as needed. Thus, PSpice is seen to be a substitute for the cut-and-try

7

method of testing circuit prototypes in the laboratory until the desired performance is achieved.

In some ways, using PSpice can be like trying to register a car by mail:

Registering a Car by Mail	Running PSpice
Fill out a form and write a check.	Create a PSpice input file.
Send the envelope off to the friendly folks at the Motor Vehicle Dept. (MVD).	Run PSpice in batch mode.
Much later, receive a large envelope from MVD, contents unknown.	See that PSpice has stopped running on the PC display meaning that PSpice has created an output file.
Open envelope, read contents.	Examine PSpice output file on PC display.

When you open the MVD envelope you might find a list of 73 reasons why you can't register the car <u>with the form the way it is</u>. The PSpice output file may contain error messages which tell you what is wrong with your input file, or perhaps the analysis was successful but the results are not presented the way you'd like to see them.

Just possibly the envelope contains your new license plates, and just possibly the circuit analysis went precisely the way you intended. In fact, you don't know whether you have been successful at either venture until you examine the results. Just as the MVD form can be changed and mailed off again, the input file can be edited and resubmitted to PSpice. Eventually the license plates will arrive and the output file will have all you want in it. Frankly, I'm confident you'll have greater success with PSpice than with most MVDs.

2.2 PROCESS FOR USING PSpice

The process for using PSpice to analyze an electronic circuit is detailed below. Figure 2.1 is a flowchart showing the 4 necessary steps.

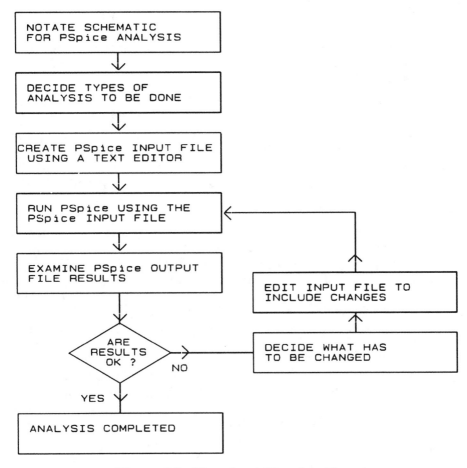

Figure 2.1 Flowchart Showing How to Analyze a Circuit with PSpice.

1. Start with a schematic diagram of the circuit and notate it for PSpice. Notation consists of three steps:

   a. Give each component, or circuit element, a name. For example, a 10K ohm resistor could be named R, R7, RLOAD or RBASEQ1. Notice that the name of a resistor must start with the letter R, and can contain from 1 to 8 characters. Each element must have a different name. Capacitors start with C, inductors begin with L, etc.

   b. Assign one node (a point of connection between two or more circuit elements) the node number zero. This is the ground (or datum) node, and all other circuit node voltages will be expressed with respect to this node. While the datum node need not be the actual circuit ground node, it is sensible to assign it this way.

c. Each additional node in the circuit is given a node number (or letter or name, such as OUTPUT). Numbers must be positive integers. The order in which nodes are numbered is arbitrary, and node numbers need not be sequential. Each node must be connected to at least two elements, except for transmission lines and MOSFET substrates.

2. Decide what type(s) of analysis you want to perform on the circuit. PSpice can do DC, AC, transient, DC transfer function, DC small-signal sensitivity, noise and Fourier analyses. Based on the types of analyses to be done, one or more control lines will have to be added to the input file. A control line could specify the values of DC voltage or current for a source, the range of frequencies an AC source is to have, or the time interval over which a transient analysis is to take place and the time step size to be used. Control lines also instruct PSpice how to present the results of the analyses; tabular form and graphical form are available. Probe, a graphics post-processor, can provide high-quality graphs of analysis results.

3. Create an input file for PSpice using a text editor. The first line of the input file MUST be a title line. The last line of the input file MUST be .END with nothing following it. Note that in the .END line the period (.) is a necessary part of the line. Other than the title line and the .END line, the element lines and the control lines can be in any order (one exception to this is when subcircuits are used). Generic and annotated sample PSpice input files are shown in Figs. 2.2 & 2.3, respectively. Additional sample PSpice input files can be found in Chaps. 3, 6, 7, 9, 10 and 12 as well as in Appendices C and E.

4. Run PSpice, which will perform a circuit analysis and create an output file. Examine the output file. If the results are not satisfactory, revise the circuit as necessary, edit the input file to reflect the change, and run PSpice again.

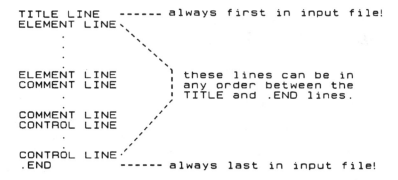

```
TITLE LINE     ------ always first in input file!
ELEMENT LINE
        .
        .
        .
ELEMENT LINE                these lines can be in
COMMENT LINE                any order between the
        .                   TITLE and .END lines.
COMMENT LINE
CONTROL LINE
        .
        .
CONTROL LINE
.END           ------ always last in input file!
```

Figure 2.2 Generic PSpice Input File.

ANNOTATED INPUT FILE				*This title line is a must!*
VBattery		**11**	**0**	**6**
* Battery is a NI-CAD type				
RA		**11**	**6**	**20**
RC		**6**	**0**	**51**
RB		**6**	**18**	**47**
* Remember R Tolerance!				
LOUTPUT		**18**	**0**	**3**
.OP				
.END				

ANNOTATED INPUT FILE — *This title line is a must!*
VBattery   11 0 6 — *A DC source, 6 V, node 11 is +.*
* Battery is a NI-CAD type — *A comment line reminds you later.*
RA   11 6 20 — *Element line for 20 ohm resistor.*
RC   6   0 51 — *Here's a 51 ohm resistor.*
RB   6 18 47 — *47 ohm R between nodes 6 and 18.*
* Remember R Tolerance! — *Comments make file readable.*
LOUTPUT   18 0 3 — *Element line for 3 H inductor.*
.OP — *Control line; does operating point.*
.END — *This line must be the last one.*

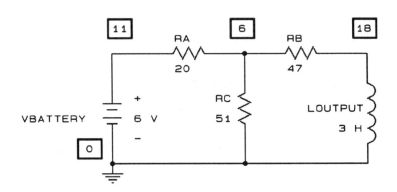

**Figure 2.3 (a) Annotated Sample PSpice
Input File; (b) Circuit Described by
Sample PSpice Input File.**

2.3 RIGID RULES, HANDY HINTS AND USEFUL THINGS TO REMEMBER

1. Each node must have a DC path to ground, which is node zero. This is necessary so that PSpice can do an initial DC small-signal analysis to determine the DC voltage at each node. The simple case where two capacitors are in series, with a node at their junction, will lead to an error message and the aborting of the PSpice analysis. One way to avoid this is to place a large value of resistor (1 giga-ohm or 1 tera-ohm) in parallel with a capacitor, thus providing a DC path to ground. Such a large value of resistance will not materially alter the circuit performance, and will allow PSpice to run the analysis.

2. PSpice will not allow a loop of voltage sources and/or inductors, nor a series connection of current sources and/or capacitors. The use of a very small resistor in series with an inductor, or a very large resistor in parallel with a capacitor (see hint 1 above) will eliminate this problem.

3. Comment lines are very useful in programs that are to be looked at by a human being, even when that human is the author of the programs. The same is true of comments in PSpice input files. Well-crafted comments will help when one refers back to an old input file to assist in creating a new file.

To make a comment line, put an asterisk (*) in the first column and write the comment. Comments may appear anywhere in the input file except for the first line (which is always the title line) and the last line (which is always .END). PSpice ignores all comment lines; however, you will be glad you took the time to put them in.

4. Creating the .END line may get you into trouble. Some text editors will put in a blank, empty line after the last carriage return (or ENTER). If you type .END and then hit ENTER, an empty, blank line (but nonetheless a line) is put after the .END line. When PSpice examines the input file, it will find that the last line is not .END and may give you an error message: "WARNING - there are no devices in this circuit." If you get this disconcerting message, carefully examine the input file to make sure that the very last line is .END and not a blank.

5. Some text editors put "funny" characters in what to you looks like a plain text file. When PSpice examines the

input file and encounters these "funny" characters, errors can result. Be sure the text editor you use to create input files is in non-document mode and does not add undesired characters.

6. PSpice will recognize numbers expressed in a variety of ways. Numbers can be integers (-12, 374), floating point (2.123, -54.997), in scientific notation (8.623E2, -3E04, 10.775E-5, where E denotes the power of 10), or can use the following engineering-notation suffixes:

F = 1E-15 (femto)	K = 1E3	(Kilo)
P = 1E-12 (pico)	MEG = 1E6	(Mega)
N = 1E-9 (nano)	G = 1E9	(Giga)
U = 1E-6 (micro)	T = 1E12	(Tera)
M = 1E-3 (milli)	MIL = 25.4E-6	

So, 270 may be expressed as 270, 270.0, 2.7E2, or 0.27K. Also, 0.000543 could be represented by 5.43E-4, 0.543M, and 543U. Any letters following one of the suffixes above will be ignored, as will letters that are not suffixes following a number. Thus, 270V, 2.70E2VOLTS, 0.27KOHM, and .27KW all mean the same number, 270; the letters which follow valid numbers (VOLTS, OHM, W) are without meaning to PSpice.

It is important to note here that M, which we associate normally with Mega, in PSpice means milli. Mega must be denoted by MEG.

At this point, you may want to glance quickly at the rest of this chapter and then jump ahead to Chap. 3 to see some representative circuits, their PSpice input files, and the resulting PSpice output files. The remainder of this chapter is a detailed presentation of element lines, to which you can return as needed to understand subsequent examples or when writing your own input files.

2.4 RULES FOR ELEMENT LINES (R, C, L & K)

In the sample element lines which follow, XXXXXX, YYYYYY, and ZZZZZZ mean that any alphanumeric (including "$", "_", "*", "/" and "%") string totaling up to 131 characters can be used after the letter (R, C, L, etc.) which specifies the type of element. It is wise to limit element names to eight characters total. Examples of valid names for a capacitor are C, CFILTER, C472, and CINPUT34. Any data contained within a less-than sign (<) and a greater-than sign (>) represents optional information. Any punctuation, such as equal signs, parentheses and commas, must be included as shown. Extra spaces in an element line are ignored. An element line may be continued by entering a + (plus) sign in the first column of the next line. PSpice will start reading the continued element line in the second column.

PSpice recognizes an element by the first letter of the element name on the element line. There are 18 letters for element names as follows:

B	GaAsFET
C	Capacitor
D	Diode
E	Voltage-Controlled Voltage Source
F	Current-Controlled Current Source
G	Voltage-Controlled Current Source
H	Current-Controlled Voltage Source
I	Independent Current Source
J	Junction Field-Effect Transistor (JFET)
K	Coefficient of Coupling, for Mutual Inductance
L	Inductor
M	Metal-Oxide Semiconductor Field-Effect Transistor (MOSFET)
Q	Bipolar Junction Transistor (BJT)
R	Resistor
S	Voltage-controlled Switch
T	Transmission Line
V	Independent Voltage Source
W	Current-controlled Switch

In this chapter, resistor, capacitor, inductor, mutual inductor, independent current and voltage sources, and switches will be presented. These basic circuit elements will allow you to do analyses on a vast array of circuit types. Controlled (or dependent) current and voltage sources are discussed in Chap. 5; semiconductors (diode, BJT, GaAsFET, JFET and MOSFET) will be covered in Chap. 8; transmission lines are presented in Chap. 10.

Note: MicroSim Corporation offers a Digital Simulation Option to PSpice, which includes the following elements (U - digital device, N - digital input, and O - digital output). For information on using the Digital Simulation Option, refer to the PSpice User's Guide.

2.4.1 Resistors

RXXXXXX N1 N2 VALUE <TC=TC1<,TC2>>

TC1 and TC2 are optional temperature coefficients; if not specified, they are both assumed to be zero. The value of the resistor as a function of temperature is given by:

$$VALUE(temp) = VALUE(tnom)*(1+TC1(temp-tnom) + TC2*(temp-tnom)^2)$$

EXAMPLES:

RINPUT 8 17 3.5E2 TC = -0.004, 0

This describes a resistor between nodes 8 and 17, with a negative and linear temperature coefficient. At the nominal temperature, its resistance is 350 ohms. If the nominal temperature is 27 C, at 127 C the resistance would be given by:

VALUE = 350 * (1 + (-0.004)(127 - 27))
VALUE = 350 * (0.6) = 210 ohms

R12 5 9 2700K

A 2.7 megohm resistor between nodes 5 and 9.

RWIRE 6 31 0.008

An 8 milli-ohm resistor between nodes 6 and 31.

Note: resistors can also have a Model Name between N2 and Value above; if so, a .MODEL line for the resistor model must be present. See Appendix A.7 and A.8 for examples.

2.4.2 Capacitors

CYYYYYY N+ N- VALUE <IC=INCOND>

N+ and N- are the nodes to which the capacitor is connected; N+ is the positive node and N- is the negative node. IC is the value of voltage across the capacitor at time zero and is optional. If an initial condition is given in the element line for a capacitor, it applies only if the UIC (use initial conditions) option is included on the .TRAN control line. .TRAN causes a transient analysis to be done; this is covered in Section 5.4.

EXAMPLES:

CBYPASS 3 0 4.7E-8

This describes a 0.047 microfarad capacitor between nodes 3 and 0.

CFILTER 14 22 33U IC=12V

This is a 33 microfarad capacitor, connected between nodes 14 and 22, initially charged (at time = 0) to +12 volts DC (node 14 positive compared to node 22).

C3 35 8 9100P

A 9.1 nanofarad capacitor between nodes 35 and 8.

A nonlinear capacitor whose capacitance depends on the instantaneous voltage across it can be described to PSpice by

CXXXXXX N+ N- POLY C0 C1 C2 . . . <IC = INCOND>

where C0, C1, C2 . . . are coefficients of a polynomial which determines the capacitance. The expression for the capacitance (in farads) is

$$C = C0 + C1*V + C2*V^2 + . . .$$

where V is the instantaneous voltage across the capacitor.

Note: capacitors can also have a Model Name between N- and Value above; if so, a .MODEL line for the capacitor model must be present. See Appendix A.7 and A.8 for examples.

2.4.3 Inductors

LZZZZZZ N+ N- VALUE <IC=INCOND>

N+ and N- are the nodes to which the inductor is connected; N+ is the positive node and N- is the negative node. IC is the value of current flowing through the inductor at time zero, and is optional. Positive current is assumed to flow from the N+ node, through the inductor, to the N- node. If an initial condition is given in the element line for an inductor, it applies only if the UIC (use initial conditions) option is included on the .TRAN control line.

EXAMPLES:

LTANK 11 34 5.6E-4

A 560 microhenry inductor between nodes 11 and 34.

LOSC 2 55 12M

A 12 millihenry inductor between nodes 2 and 55.

L7 29 41 2.85 IC = 3

This describes a 2.85 henry inductor between nodes 29 and 41, with a current at time zero of 3 amps flowing from node 29 to node 41 through the inductor.

A nonlinear inductor whose inductance depends on the instantaneous current through it can be described as follows:

LXXXXXX N+ N- POLY L0 L1 L2 . . . <IC = INCOND>

where L0, L1, L2 . . . are coefficients of a polynomial which determines the inductance. The expression for the inductance (in henries) is

$$L = L0 + L1*I + L2*I^2 + . . .$$

where I is the instantaneous current through the inductor.

Note: inductors can also have a Model Name between N- and Value above; if so, a .MODEL line for the inductor model must be present. See Appendix A.7 and A.8 for examples.

2.4.4 Mutual Inductors

KXXXXXX LYYYYYY LZZZZZZ VALUE

This is an element line which allows two inductors (each of which is described by its own element line) to be magnetically linked. VALUE is the magnitude of the coefficient of coupling, which must be greater than 0 and must not exceed 1. Notice that this element line has no node numbers, since the two inductors stated in this line each have their own node connections. The dotted ends of the inductors are assumed by PSpice to be the N+ nodes of each.

EXAMPLES:

KPRISEC LP LS 0.97

   The coefficient of coupling between LP and LS is 0.97.

KAX LANT LXMTR .08

   The coefficient of coupling between LANT and LXMTR is 0.08.

2.5 INDEPENDENT SOURCES

Vxxxxxx N+ N-< <<DC> DC/TRAN-VALUE> <AC <ACMAG <ACPHASE>>>
            <TRANKIND(TPAR1 TPAR2 . . .)> >

Ixxxxxx N+ N- < <<DC> DC/TRAN-VALUE> <AC <ACMAG <ACPHASE>>>
            <TRANKIND(TPAR1 TPAR2 . . .)> >

The word "independent" in independent source means that the current flowing through the current source, or the voltage across the voltage source, is not dependent on any other circuit parameter. Examples of independent sources are ideal batteries, an electric power outlet (for reasonable load currents), or a function generator with zero output impedance.
   The collector-emitter current in a common-emitter transistor circuit can be thought of as a "dependent" source, since the C-E current is controlled by, or is dependent upon, the base current. For a typical small-signal transistor, collector current would be 100 times the base current. This would be a current-controlled current source. A field-effect transistor (FET) would be an example of a voltage-controlled current source, since the drain-source current is controlled by the gate-source voltage.
   The letters DC following the N- node number indicate that the source does not vary with time, and are optional. DC/TRAN-VALUE is the magnitude of the source for DC and transient analysis; this value may be left out if it is zero. The letters AC indicate that the source is a sinusoid with magnitude and phase described by ACMAG and ACPHASE when an AC

analysis is performed. ACMAG and/or ACPHASE may be omitted; ACMAG will default to one volt, while ACPHASE will default to zero degrees.

In addition, for the purpose of transient analysis only, an independent source can be one of five time-dependent functions (TRANKIND): pulse, sinusoidal, exponential, piece-wise linear or single-frequency FM. TPAR1 etc. are the parameters associated with the particular TRANKIND. These five time-dependent functions are described in detail later in the chapter.

PSpice considers positive current to be that current which flows from the positive node (N+) to the negative node (N-) through the source. When the product of voltage and current is positive, a source is receiving power from the circuit; if the product is negative, the source is supplying power to the circuit.

While it is possible to have all the options specified for an independent source (DC, AC and TRANKIND), it would be better in most cases for reasons of clarity (and sanity) to use only one option at a time with a source. It is also possible to specify none of the options, in which case the source is dead. A dead voltage source, which acts like a short circuit, has a very useful purpose. See the VBORING example below.

EXAMPLES:

VBATTERY 12 23 DC 6

This is a 6 V DC voltage source, with node 12 positive and node 23 negative.

ILED 38 46 SIN(20M 80M 10KHZ)

This describes a current source with a DC value (or offset) of 20 mA, and a time-dependent value (when a transient analysis is performed) of a 10 KHz sinusoid with a peak amplitude of 80 mA. Positive current is assumed to flow from node 38 to node 46, through the current source.

VBORING 7 5

VBORING is a dead voltage source between nodes 7 and 5; it always has zero volts across it, and behaves like a short circuit. While it may appear to be useless, it can be used as an ammeter to measure currents in any circuit branch desired. PSpice uses dead voltage sources for this purpose.   Note: if the current in VBORING were positive, that would mean current was flowing from node 7 to node 5 through VBORING.

VFMSIG 29 11 AC 1 45 SFFM(0 2 1K 2.5 200)

The voltage source VFMSIG can be either of two things, depending on what type of analysis is being done. When an AC analysis is done (by a separate control line starting with .AC), VFMSIG is an AC sinusoidal source with a magnitude of 1 V and a phase angle of 45 degrees. The frequency is set by the .AC control line. When a transient analysis is performed (by a separate control line starting with .TRAN), VFMSIG becomes a single-frequency FM source, with an amplitude of 2 V peak, carrier frequency of 1 KHz, modulation index of 2.5, and modulating (or signal) frequency of 200 Hz.

This example shows a two-purpose source that can be described to PSpice with a single element line. In many cases it would be better to run PSpice twice: first with the AC source and AC analysis, and then with the single-frequency FM source and transient analysis. To do that, a text editor would be used to edit the input files so that one contained VFMSIG 29 11 AC 1 45 and the other contained VFMSIG 29 11 SFFM(0 2 1K 2.5 200).

VSYNC 8 17 PULSE(-10 20 10M 1U 2U 15M 60M)

VSYNC is a pulse train starting at -10 V, going to 20 V, delayed from time-zero by 10 ms, with a rise time of 1 microsec, fall time of 2 microsec, pulse width of 15 ms, and period of 60 ms.

ITRIANGL 22 34 PWL(0 0 .25 3 .75 -3 1.25 3 1.75 -3 2 0)

This describes two cycles of a triangular current waveform, starting at 0 V and ramping between +3 V and -3 V with a period of 1 s. Positive current flows from node 22 to node 34 through the current source ITRIANGL.

2.5.1 Time-Dependent Functions for Independent Sources

Five kinds of time-dependent functions are available for transient analysis using independent sources:

1. PULSE - A single pulse or pulse train, with rise and fall times that can be specified. This can be used to produce pulse, squarewave, triangle and sawtooth sources.

2. SIN - A sinusoid in which the start can occur after time zero. It may be exponentially damped if desired.

3. EXP - A single pulse, with exponentially growing rise and exponentially decaying fall; rise and fall may have different time constants.

4. PWL - A piece-wise linear description of a source which can be used to approximate nearly any waveform (if you have the time to generate the time-voltage coordinate list). For example, a complex TV sync waveform or an electrocardiogram can be described to PSpice using PWL.

5. SFFM - Single-frequency FM (frequency modulation) source with independently-specified offset voltage, amplitude, carrier frequency, FM modulation index and modulation frequency (limited to a single sinusoid).

2.5.1.1 Pulse Function for Independent Sources

```
VXXXXXX    N+ N- PULSE(V1 V2 TD TR TF PW PER)
IXXXXXX    N+ N- PULSE(V1 V2 TD TR TF PW PER)
```

describe a voltage or current pulse source. The parameters are:

Parameter	Default Value	Unit
V1 (initial value)	must specify	Volts or amps
V2 (pulsed value)	must specify	Volts or amps
TD (time delay)	0.0	seconds
TR (rise time)	TSTEP	seconds
TF (fall time)	TSTEP	seconds
PW (pulse width)	TSTOP	seconds
PER (period)	TSTOP	seconds

TSTEP and TSTOP refer to the step size and stop time in the transient analysis which is specified by a .TRAN TSTEP TSTOP control line in the input file. A pulse waveform illustrating the parameters above is shown in Fig. 2.4.

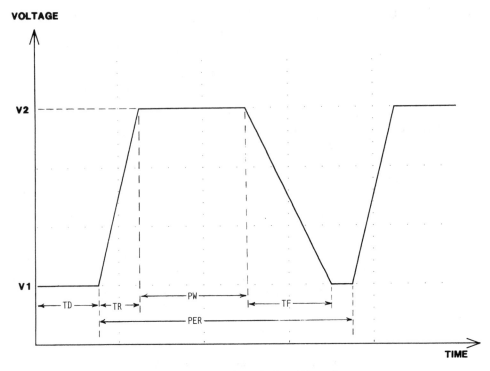

Figure 2.4 PSpice Pulse Waveform.

2.5.1.2 Sinusoidal Function for Independent Sources

VXXXXXX   N+ N- SIN(VO VA FREQ TD DF PHASE)
IXXXXXX   N+ N- SIN(VO VA FREQ TD DF PHASE)

describe a voltage or current sinusoidal source. The parameters are:

Parameter		Default Value	Unit
VO	(offset value)	must specify	Volts or amps
VA	(amplitude)	must specify	Volts or amps
FREQ	(frequency)	1/TSTOP	Hertz
TD	(delay time)	0.0	second
DF	(damping factor)	0.0	1/second
PHASE	(phase angle)	0.0	degree

TSTOP refers to the stop time in the transient analysis which is specified by a .TRAN TSTEP TSTOP control line in the input file. A damped sinusoidal waveform illustrating the parameters above is shown in Fig. 2.5.

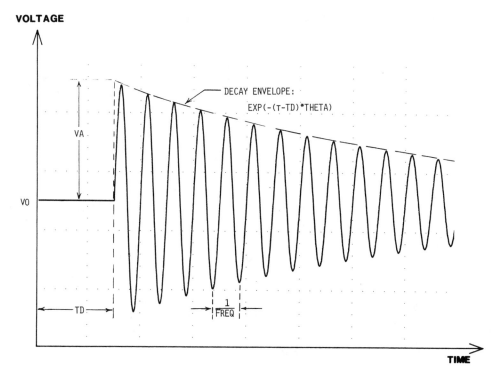

VOLTAGE

DECAY ENVELOPE:
EXP(-(T-TD)*THETA)

VA

VO

TD

$\frac{1}{FREQ}$

TIME

Figure 2.5 PSpice Sinusoidal Waveform.

The waveform as a function of time, t, is described by the relationships:

from t = 0 to TD,   VXXXXXX = VO+VA*sine(2*PI*PHASE/360)
from t = TD to TSTOP,   VXXXXXX =

VO + VA*exp(-(t-TD)*DF)*sine(2*PI*FREQ*(t-TD)+PHASE/360)

If TD is not specified, the sine starts at time-zero. If DF is not specified, the sine has no decay and therefore has constant peak amplitude. If PHASE is not specified, the sinusoid starts at zero degrees.

2.5.1.3 Exponential Function for Independent Sources

VXXXXXX   N+ N- EXP(V1 V2 TD1 TAU1 TD2 TAU2)
IXXXXXX   N+ N- EXP(V1 V2 TD1 TAU1 TD2 TAU2)

describe a voltage or current exponential source. The parameters are:

Parameter		Default Value	Unit
V1	(initial value)	must specify	Volts or amps
V2	(pulsed value)	must specify	Volts or amps
TD1	(rise delay time)	0.0	seconds
TAU1	(rise time constant)	TSTEP	seconds
TD2	(fall delay time)	TD1 + TSTEP	seconds
TAU2	(fall time constant)	TSTEP	seconds

TSTEP refers to the step time in the transient analysis which is specified by a .TRAN TSTEP TSTOP control line in the input file. Figure 2.6 shows an exponential waveform illustrating the parameters above.

The waveform as a function of time, t, is described by the relationships:

from t = 0 to TD1,   VXXXXXX = V1
from t = TD1 to TD2,   VXXXXXX =

$$V1 + (V2-V1)*(1-exp(-(t-TD1)/TAU1))$$

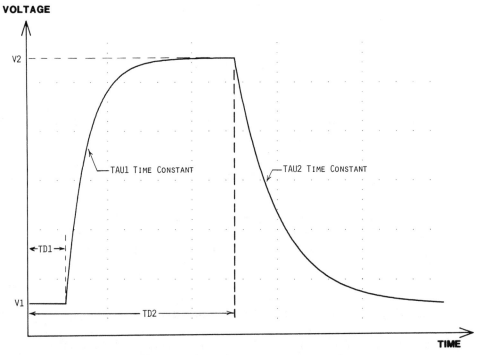

Figure 2.6 PSpice Exponential Waveform.

from t = TD2 to TSTOP,   VXXXXXX =

$$V1 + (V2-V1)*(1-\exp(-(t-TD1)/TAU1)) +$$
$$(V1-V2)*(1-\exp(-(t-TD2)/TAU2))$$

2.5.1.4 Piece-wise Linear Function for Independent Sources

```
VXXXXXX    N+ N- PWL(T1 V1  <T2 V2  T3 V3 . . .>)
IXXXXXX    N+ N- PWL(T1 V1  <T2 V2  T3 V3 . . .>)
```

describe a voltage or current piece-wise linear source. The parameters are:

Parameter	Default Value	Unit
Tn (nth time value)	none	seconds
Vn (nth voltage value)	none	Volts or amps

Any arbitrary voltage or current waveform may be described by a piece-wise linear approximation. Each time and voltage (or current) pair is the coordinate of a point on the graph of the waveform versus time. Time points need not be evenly spaced, and the source value at times between specified points is done by linear interpolation. Figure 2.7 shows a PWL voltage source described by the element lines

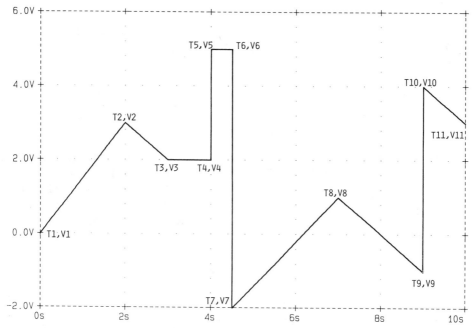

Figure 2.7 PSpice Piece-wise Linear Waveform.

```
VPWL 1 0 PWL(0 0  2 3  3 2  4 2  4.01 5
+ 4.5 5  4.51 -2  7 1  9 -1  9.01 4  10 3)
```

Notice that the element "line" really consists of two lines in
the input file; the second line contains a plus sign (+) in
the first column.   PSpice interprets this to be a continuation
of the element line immediately preceding.   In order to approximate accurately a complex waveform using a piece-wise linear
approximation, a large number of lines would have to be used.
A PWL source element "line" describing an electrocardiogram
signal every 5 ms for 500 ms uses 16 lines of the input file.
In a case such as this, it would be wise to create a text file
containing the complex PWL source by itself.   Whenever that
source was needed for a PSpice input file, the text file could
be appended to the input file, saving a great deal of tedious
typing.

2.5.1.5 Single-Frequency FM Function for Independent Sources

```
VXXXXXX    N+ N- SFFM(VO VA FC MDI FS)
IXXXXXX    N+ N- SFFM(VO VA FC MDI FS)
```

describe a voltage or current single-frequency FM (frequency
modulation) source. The parameters are:

Parameter		Default Value	Unit
VO	(offset)	must specify	Volts or amps
VA	(amplitude)	must specify	Volts or amps
FC	(carrier frequency)	1/TSTOP	Hz
MDI	(modulation index)	must specify	none
FS	(signal frequency)	1/TSTOP	Hz

TSTOP refers to the step time in the transient analysis which
is specified by a .TRAN TSTEP TSTOP control line in the input
file.   Figure 2.8 shows a single-frequency FM waveform illustrating VO and VA; it shows an unusually large amount of deviation (500 KHz with a carrier frequency of 1 MHz) and was
created by the element line

VSAMPLE 1 0 SFFM(2 4 1MEG 5 100K)

The modulation index is the deviation (in Hz) divided by the
modulating signal frequency (in Hz), and has no unit.   The

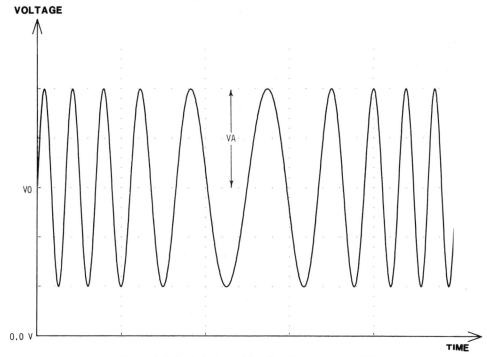

Figure 2.8 PSpice Single-Frequency FM Waveform.

waveform as a function of time, t, is described by the relationship:

VXXXXXX = VO + VA*sine((2*PI*FC*t) + MDI*sine(2*PI*FS*t))

2.6 SWITCHES

PSpice, unlike SPICE, allows voltage-controlled and current-controlled switches to be used in circuits. Such circuit elements are very useful when simulating such things as turning a load on and off, stopping the charge or discharge of a capacitor, a diode or fuse becoming an open circuit during an analysis, or the operation of a switching power supply. Switches are turned on and off by either the voltage between two nodes in the circuit, or by the current through a voltage source in the circuit, reaching a threshold value.

   Like real switches, they have definable resistances in both the ON condition and the OFF condition. Unlike real

switches, they have resistances that vary continuously between the ON-state value and the OFF-state value as controlling voltage (or current) varies between the ON threshold and
the OFF threshold.

2.6.1 Voltage-Controlled Switches

SXXXXXX N1 N2 NC+ NC- MODELNAME
.MODEL MODELNAME VSWITCH RON ROFF VON VOFF

N1 and N2 are the nodes to which the switch is connected,
NC+ and NC- are the positive and negative controlling
nodes, and MODELNAME is the name associated with the characteristics of this switch. There must be a .MODEL line
associated with the MODELNAME, although a single .MODEL
line can be used by more than one switch.

EXAMPLE:

S/FUSE 21 11 7 0 SLOBLO
.MODEL SLOBLO VSWITCH ron=.1 roff=1G von=4 voff=3

This describes a switch connected between nodes 21 and
11, controlled by the voltage between nodes 7 and 0, with
an ON resistance of 100 milliohms, and an OFF resistance
of 1000 Megohms. When the voltage between nodes 7 and 0
is equal to or above 4 V, the switch is fully ON (resistance of .1 ohms). When the voltage between nodes 7 and
0 is less than or equal to 3 V, the switch is fully OFF
(resistance of 1 gigohm). If the controlling voltage is
between 3 and 4 volts, a rather complex relationship exists between controlling voltage and resistance, and the
resistance varies continuously. This represents a normally open switch in that when the controlling voltage is
zero, the switch is off (open condition).

For any switch, it is important to keep the ratio of RON/ROFF
to 1E12 or less, to avoid exceeding the dynamic range of
PSpice. A good rule of thumb to follow when simulating ideal
switches is to make the RON negligibly small compared to circuit impedances, and to make ROFF much larger than circuit impedances. There's no need to make ROFF 1E12 ohms when 1E8
ohms would do as well. When you make a switch model for a
real device (like a FET analog switch), RON should be set to
a realistic level found in the data sheet.

Also, if VON and VOFF are very close, numerical problems can occur, leading to long analysis times. RON and ROFF both must be larger than 0 and less than 1/GMIN, where GMIN is the minimum conductance allowed by PSpice. The default value for GMIN is 1E-12, but can be altered by using the .OPTIONS control line.

2.6.2 Current-Controlled Switches

WXXXXXX N1 N2 VNAME MODELNAME
.MODEL MODELNAME ISWITCH RON ROFF ION IOFF

N1 and N2 are the nodes to which the switch is connected, VNAME is the voltage source through which the controlling current flows, and MODELNAME is the name associated with the characteristics of this switch. There must be a .MODEL line associated with the MODELNAME, although a single .MODEL line can be used by more than one switch.

EXAMPLE:

W/RELAY 3 8 VCOIL PWR
.MODEL PWR ISWITCH ron=2 roff=10MEG ion=5M ioff=9M

This describes a switch connected between nodes 3 and 8, controlled by the current through voltage source VCOIL (located elsewhere in the circuit), with an ON resistance of 2 ohms, and an OFF resistance of 10 Megohms. When the current through VCOIL is less than or equal to 5 mA, the switch is fully ON (resistance of 2 ohms). When the current through VCOIL is greater than or equal to 9 mA, the switch is fully OFF (resistance of 10 megohm). If the controlling current is between 9 mA and 5 mA, a rather complex relationship exists between controlling current and resistance, and the resistance varies continuously. This represents a normally closed switch in that when the controlling current is zero, the switch is on (closed condition).

CHAPTER SUMMARY

PSpice is not an interactive program. You may have to submit an edited input file several times to get the results looking the way you want them.

Each node in a PSpice input file must have a DC path to node zero (ground). You may have to put in dummy resistors of very large value (so as not to disturb circuit operation) to satisfy this requirement.

Comments are very handy additions to PSpice input files.

Integer, floating point, scientific notation and engineering notation numbers are all usable in input files.

Element names should have 8 or fewer characters.

Independent current and voltage sources can have DC and AC values, and can be the following types of time-dependent functions: pulse, sinusoidal, exponential, piece-wise linear and single-frequency FM. It is best not to give a source DC, AC and time-dependent values, if only to avoid confusion. You can always run two or three different analyses, with the source edited to have only one value at a time.

Switches can be simulated by PSpice. Voltage- and current-controlled switches may be normally open or normally closed, and can have their resistances in the ON and OFF states defined.

Chapter 3

FIRST ATTEMPTS
AT ANALYSIS,
WITH SAMPLE CIRCUITS

3.1 INTRODUCTION

In Chap. 2 some of the ground rules of using PSpice were covered. In this chapter, a variety of examples will be presented to illustrate what an input file looks like and what the
output file created by PSpice analysis contains. By looking
at three things, the circuit schematic diagram, the input
file and the output file, you will gain an appreciation for
what PSpice can do and you will start developing the ability
to use PSpice yourself. For a detailed explanation of each
PSpice feature illustrated, refer to the chapter in which
that feature is covered. A good way to begin learning PSpice
is to create input files similar to the ones in this chapter,
and run analyses of them. *Note: Probe, a powerful graphics
post-processor which takes analysis results and makes graphs,
is not covered here; it will be introduced in Chap. 7.*

3.2 THE FIRST CIRCUIT - 3 RESISTORS, 1 BATTERY

Let's begin with a DC circuit, containing a battery (independent voltage source) and three resistors. The circuit dia-

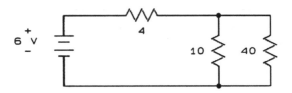

Figure 3.1 3 Resistor Circuit.

gram is shown in Fig. 3.1. A convenient node to choose for the datum node is the negative side of the battery; this is labeled node 0. The two other nodes which must be labeled are the positive side of the battery and the junction of the three resistors. Since the only restriction on node numbers is that they not be negative integers, let the two other nodes be numbered 17 and 84 (any other alphanumeric strings would do just fine!). The notated circuit is shown in Fig. 3.2.

Any input file for PSpice analysis <u>must</u> contain the following:

1. a title line

2. element lines for each element (R, L, C, etc.) in the circuit

3. a .END line. A period (.) must appear before END.

Other lines, which are optional, include control lines (which cause things to happen like changing the voltage or frequency of a source, setting the temperature(s) of the circuit, specifying the type(s) of analysis to be done, printing out a graph or table of results, etc.) and comment lines, which make understanding the input file much easier weeks or months later.

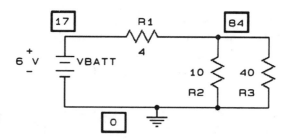

Figure 3.2 3 Resistor Circuit Notated for PSpice.

First Attempts at Analysis, with Sample Circuits   Chap. 3

```
3R.CIR   THREE RESISTOR, ONE BATTERY PROBLEM
VBATT    17  0  6
R1       17 84  4
R2       84  0 10
R3       84  0 40
.END
```

**Figure 3.3 PSpice Input File for 3
Resistor Circuit.**

Shown in Fig. 3.3, the input file for the circuit of Fig.
3.2 contains only a title line, four element lines and a .END
line. Other than the title line and the .END line, the order
of all other lines is arbitrary. Notice that each element
has a name as well as a value. For example, the 4 ohm resistor is named R1. The names are arbitrary as well, except
that all resistors must begin with an R, all voltage sources
must begin with a V, etc.

Although no specific analysis is requested, PSpice automatically does a DC analysis, called a "Small Signal Bias Solution," and prints the results in the output file. The title
line can contain anything whatsoever; however, in the interests of organization and preservation of one's sanity, it is
best to make the title contain two things:

1. the name of the input file. In this example, 3R.CIR
is the name of the file containing the title line, the
element lines and the .END line. Doing this makes it
easier to find an input file in the directory of your computer at a later time; there will be many input files
when you become a regular user of PSpice.

2. a brief description of what the circuit is (THREE RES-
ISTOR, ONE BATTERY PROBLEM). This is useful because both
the input file and the output file lack a schematic diagram of the circuit being analyzed. Though one can study
an input file and thereby reconstruct the circuit so described, it helps to have a clue in the title line, since
the title line normally will be printed on each page of
the output file.

The output file, shown in Fig. 3.4, contains two parts: the
first is a listing of the input file, called "Circuit Description," and the second is the "Small Signal Bias Solution,"
which gives the DC voltage at each node other than the datum
(0) node. Notice that the voltage at node 17 is 6.000 VDC
(the battery voltage) and the voltage at node 84 is 4.000

```
**** 08/02/91 17:11:14 ********* Evaluation PSpice (January 1991) ************
3R.CIR   THREE RESISTOR, ONE BATTERY PROBLEM

 ****      CIRCUIT DESCRIPTION
****************************************************************************

VBATT   17  0  6
R1      17  84  4
R2      84  0  10
R3      84  0  40
.END


**** 08/02/91 17:11:14 ********* Evaluation PSpice (January 1991) ************
3R.CIR   THREE RESISTOR, ONE BATTERY PROBLEM

 ****      SMALL SIGNAL BIAS SOLUTION        TEMPERATURE =   27.000 DEG C
****************************************************************************

 NODE    VOLTAGE      NODE   VOLTAGE      NODE   VOLTAGE      NODE   VOLTAGE
(  17)    6.0000   (  84)    4.0000


    VOLTAGE SOURCE CURRENTS
    NAME          CURRENT
    VBATT        -5.000E-01
    TOTAL POWER DISSIPATION    3.00E+00  WATTS


       JOB CONCLUDED
       TOTAL JOB TIME           1.53
```

**Figure 3.4 PSpice Output File for 3
Resistor Circuit.**

VDC, both compared to the datum node.   All node voltages
produced by the Small Signal Bias Solution are with respect
to the datum node; thus, the voltage at node 0 is, by
definition, zero volts.

3.3 HOW TO RUN PSpice
ON A PERSONAL COMPUTER

In order to perform a SPICE analysis, a version of the SPICE
program must be present on the particular computer system one
is using.   The steps that follow show how PSpice is run on a
personal computer (assuming that PSpice has been installed on
the PC according to the directions in Section 1.4).

What you must type in is shown <u>underlined</u> in the steps
below:

1. Once the computer is booted, with the DOS prompt showing, go to the PSpice subdirectory.   Assuming PSpice is
on the C drive in a subdirectory called PSpice, the command to do this would be: <u>CD PSPICE</u> and hit ENTER.

2. Create an input file describing the circuit and the desired analyses. Use a text editor to do this. Call this input file FILENAME.CIR (FILENAME represents what you choose to name the file; 3R, DIFFAMP, RC-AMP, etc. are typical names). PSpice is programmed to work with input files that have the .CIR extension.

3. Next, type in <u>PSPICE FILENAME</u> and hit ENTER. Remember, FILENAME represents the name of *your* file.

4. PSPice will read the input file, check for errors, and proceed to perform the analyses stated in the input file. You will see a PSpice screen, which tells the name of the input file, the output file, and what part of the analysis is being done.

Analysis results (or one or more error messages) will be written to an output file called FILENAME.OUT. You will be aware that the analysis is over when the DOS prompt symbol reappears.

5. To look at the output file, type in <u>TYPE FILENAME.OUT</u> and hit ENTER. The output file will scroll past on the CRT monitor. A better way is to use a text editor to view the output file.

6. To obtain hardcopy of the output file (this assumes you have a printer), type in <u>PRINT FILENAME.OUT</u> and hit ENTER. The output file will be printed on your computer's printer.

A sample run of PSPice on a PC system, using the circuit (Fig. 3.1) and its input file 3R.CIR (Fig. 3.3), would look like this on the computer monitor (What you must type in is shown <u>underlined</u> in the steps below):

C:\PSPICE><u>PSPICE 3R</u> (hit ENTER) *PSpice screen appears*
C:\PSPICE><u>TYPE 3R.OUT</u> (hit ENTER) *3R.OUT scrolls by*
C:\PSPICE><u>PRINT 3R.OUT</u> (hit ENTER) *3R.OUT is printed*

One must remember that PSPice, when used this way, is not an interactive program; you have to look at the output file to see the results of the analysis. If the results are not what is desired, or if the analysis is to be run again with a parameter changed (e.g., substituting a different value of a resistor), a copy can be made of the original input file. Then the one or two lines in the copy of the input file can easily be edited to make the desired change, and the procedure in steps 3 through 6 would be followed again.

In order to gain some proficiency in using SPICE, take a simple circuit (Fig. 3.1 or similar), notate it for SPICE analysis, write an input file describing the circuit, run SPICE, and look at the output file. If you make an error in the input file, SPICE will write meaningful error messages in the output file which will help you to find and correct the error.

3.4 HOW TO AVOID WASTING PAPER

When SPICE creates the output file, it puts a header at the top of each page and puts each part of the output file on a separate page. The header contains the date the analysis was done, the version of SPICE being used and its release date, the time the analysis was done, the title line from the input file, and a lot of asterisks (**). This can result in a very large amount of paper being used for certain types of analyses. While this may be useful for some applications, for the user who wants to print the output file but conserve paper there is an easy fix.

Simply put the control line .OPTIONS NOPAGE (with no spaces in NOPAGE) in the input file, anywhere between the title line (which is always the first line) and the .END line (which is always the last line). This control line will cause page ejects to be suppressed, concatenating the printout and saving considerable amounts of paper. Figure 3.5 shows the same input file as Fig. 3.3, except that the control line .OPTIONS NOPAGE has been added. Note that this file was made by copying 3R.CIR to 3RNOPAGE.CIR and editing 3RNOPAGE.CIR. Editing here consisted of adding one control line and one comment line to the input file (and changing the title line).

The resulting output file (Fig. 3.6) is seen to have the same information as 3R.OUT, except that only one header appears and less paper is used.

```
3RNOPAGE.CIR  THREE RESISTOR, ONE BATTERY PROBLEM
VBATT   17  0   6
R1      17  84  4
R2      84  0   10
R3      84  0   40
*       THE LINE BELOW SAVES PAPER
.OPTIONS NOPAGE
.END
```

Figure 3.5 Input File 3RNOPAGE.CIR with .OPTIONS NOPAGE Added.

First Attempts at Analysis, with Sample Circuits   Chap. 3

```
**** 08/03/91 11:19:16 ********* Evaluation PSpice (January 1991) ************
 3RNOPAGE.CIR  THREE RESISTOR, ONE BATTERY PROBLEM

 ****      CIRCUIT DESCRIPTION
 *****************************************************************************

 VBATT   17  0  6
 R1      17 84  4
 R2      84  0 10
 R3      84  0 40
 *      THE LINE BELOW SAVES PAPER
 .OPTIONS NOPAGE
 .END

 ****      SMALL SIGNAL BIAS SOLUTION      TEMPERATURE =   27.000 DEG C

 NODE   VOLTAGE      NODE  VOLTAGE     NODE   VOLTAGE     NODE   VOLTAGE
 (  17)   6.0000  (   84)   4.0000

     VOLTAGE SOURCE CURRENTS
     NAME          CURRENT

     VBATT        -5.000E-01

     TOTAL POWER DISSIPATION   3.00E+00  WATTS

         JOB CONCLUDED

     TOTAL JOB TIME            1.54
```

Figure 3.6 Output File 3RNOPAGE.OUT.

3.5   THE SECOND CIRCUIT - THREE RESISTORS, TWO BATTERIES

In this DC circuit, there are two batteries. See Fig. 3.7. SPICE can be used to give a Small Signal Bias Solution and thus determine the voltage at each node. In addition to solving the circuit, you might want to know what new value for the 8 V source would make the voltage across the 3K ohm resistor be 0 V? In order to solve this with SPICE, we shall make

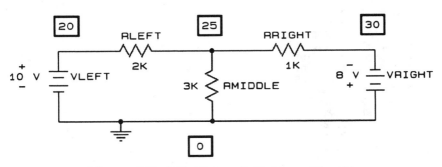

Figure 3.7 3 Resistor, 2 Battery Circuit.

```
3R2BAT.CIR    VARY ONE BATTERY VOLTAGE
VLEFT        20  0  10
RLEFT        20 25  2K
RMIDDLE      25  0  3K
RRIGHT       25 30  1K
VRIGHT        0 30   8
.DC VRIGHT 0 10 0.5
*     THE LINE ABOVE SWEEPS VRIGHT FROM 0V TO 10V IN 0.5V STEPS
.PRINT DC V(25)
.PLOT  DC V(25)
.END
```

Figure 3.8 Input File 3R2BAT.CIR.


```
**** 08/03/91 11:55:43 ********* Evaluation PSpice (January 1991) ************
 3R2BAT.CIR    VARY ONE BATTERY VOLTAGE

 ****       CIRCUIT DESCRIPTION
******************************************************************************

 VLEFT        20  0  10
 RLEFT        20 25  2K
 RMIDDLE      25  0  3K
 RRIGHT       25 30  1K
 VRIGHT        0 30   8
.DC VRIGHT 0 10 0.5
 *     THE LINE ABOVE SWEEPS VRIGHT FROM 0V TO 10V IN 0.5V STEPS
.PRINT DC V(25)
.PLOT  DC V(25)
.END


**** 08/03/91 11:55:43 ********* Evaluation PSpice (January 1991) ************
 3R2BAT.CIR    VARY ONE BATTERY VOLTAGE

 ****       DC TRANSFER CURVES              TEMPERATURE =   27.000 DEG C
******************************************************************************
   VRIGHT        V(25)
   0.000E+00    2.727E+00
   5.000E-01    2.455E+00
   1.000E+00    2.182E+00
   1.500E+00    1.909E+00
   2.000E+00    1.636E+00
   2.500E+00    1.364E+00
   3.000E+00    1.091E+00
   3.500E+00    8.182E-01
   4.000E+00    5.455E-01
   4.500E+00    2.727E-01
   5.000E+00    0.000E+00
   5.500E+00   -2.727E-01
   6.000E+00   -5.455E-01
   6.500E+00   -8.182E-01
   7.000E+00   -1.091E+00
   7.500E+00   -1.364E+00
   8.000E+00   -1.636E+00
   8.500E+00   -1.909E+00
   9.000E+00   -2.182E+00
   9.500E+00   -2.455E+00
   1.000E+01   -2.727E+00
```

Figure 3.9 Output File 3R2BAT.OUT.

First Attempts at Analysis, with Sample Circuits   Chap. 3

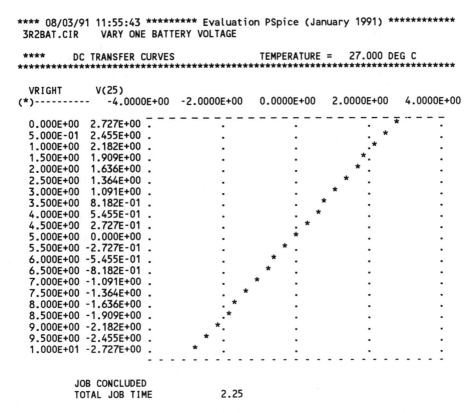

```
**** 08/03/91 11:55:43 ********* Evaluation PSpice (January 1991) ************
3R2BAT.CIR    VARY ONE BATTERY VOLTAGE

****       DC TRANSFER CURVES                 TEMPERATURE =   27.000 DEG C
*****************************************************************************

 VRIGHT        V(25)
 (*)---------    -4.0000E+00  -2.0000E+00   0.0000E+00   2.0000E+00   4.0000E+00
              - - - - - - - - - - - - - - - - - - - - - - - - - - - -
 0.000E+00  2.727E+00 .              .            .            .        * - - .
 5.000E-01  2.455E+00 .              .            .            .        *    .
 1.000E+00  2.182E+00 .              .            .            .      .*     .
 1.500E+00  1.909E+00 .              .            .            .    *.       .
 2.000E+00  1.636E+00 .              .            .            .  * .        .
 2.500E+00  1.364E+00 .              .            .            * .          .
 3.000E+00  1.091E+00 .              .            .         *  .            .
 3.500E+00  8.182E-01 .              .            .       *    .            .
 4.000E+00  5.455E-01 .              .            .     *      .            .
 4.500E+00  2.727E-01 .              .            .   *        .            .
 5.000E+00  0.000E+00 .              .            *            .            .
 5.500E+00 -2.727E-01 .              .         *  .            .            .
 6.000E+00 -5.455E-01 .              .       *    .            .            .
 6.500E+00 -8.182E-01 .              .     *      .            .            .
 7.000E+00 -1.091E+00 .              .   *        .            .            .
 7.500E+00 -1.364E+00 .              . *          .            .            .
 8.000E+00 -1.636E+00 .            .*             .            .            .
 8.500E+00 -1.909E+00 .           .*              .            .            .
 9.000E+00 -2.182E+00 .         *.                .            .            .
 9.500E+00 -2.455E+00 .       *  .                .            .            .
 1.000E+01 -2.727E+00 .     *    .                .            .            .
              - - - - - - - - - - - - - - - - - - - - - - - - - - - -

         JOB CONCLUDED
         TOTAL JOB TIME                    2.25
```

Figure 3.9 (Continued).

the battery on the right, the 8 Volt source, vary from 0 to
10 Volts, and do a DC analysis of the circuit (DC analysis is
covered in depth in Chap. 5). The results of the analysis
will be printed in tabular form and plotted as a graph of
output voltage versus battery voltage. The input file,
3R2BAT.CIR, is shown in Fig. 3.8. The stepping of VRIGHT
from 0 V to 10 V in steps of 0.5 V is accomplished by the
control line .DC VRIGHT 0 10 0.5 . The printing of a table
of the voltage at node 25 versus VRIGHT is done by the
control line .PRINT DC V(25) . The control line .PLOT DC
V(25) makes a graph of the DC voltage at node 25 versus
VRIGHT.
    The output file, 3R2BAT.OUT, is shown in Fig. 3.9. The
first page is the Circuit Description, the next page is the
table of node 25 voltage versus VRIGHT voltage, and the last
page is a graph of node 25 voltage versus VRIGHT voltage.
There are two columns of numbers on the left side of the
graph: VRIGHT, the independent variable, and V(25), the depen-

Chap. 3   First Attempts at Analysis, with Sample Circuits   39

dent variable. The horizontal V(25) axis is labeled from -4 V to +4 V. An examination of the table or the graph shows that when VRIGHT is 8 V, the voltage at node 25 is -1.636 V; and when VRIGHT is 5 V, the voltage at node 25 is 0 volts.

3.6 THE THIRD CIRCUIT - PARALLEL RESONANT TANK, AC ANALYSIS

In the schematic diagram, Fig. 3.10, a 1 V sinusoidal voltage source in series with a 1K ohm resistor is connected across a lossy tank circuit. The resonant frequency of this low-Q tank L-C is about 10 KHz. The SPICE input file in Fig. 3.11 contains the 5 element lines for the circuit, and has a control line which causes the frequency of the voltage source to vary from 5 KHz to 15 KHz, in 20 linearly-spaced steps. This causes the AC analysis to be done at 21 frequencies -- every 500 Hz. The results of the AC analysis will appear in the

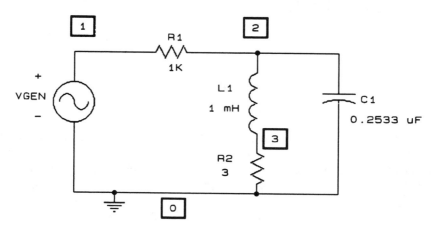

Figure 3.10 Tank Circuit.

```
ACTANK.CIR    PARALLEL RESONANT CIRCUIT, 10 KHZ
R1   1   2   1K
R2   3   0    3
VGEN 1 0 AC 1
L1   2   3   1M
C1   2   0   0.2533U
.AC LIN 21   5000 15000
.PRINT AC VM(2) VP(2)
.PLOT AC   VM(2) VP(2)
.END
```

Figure 3.11 Input File ACTANK.CIR.

```
****      CIRCUIT DESCRIPTION
*******************************************************************************

R1   1   2   1K
R2   3   0   3
VGEN 1 0 AC 1
L1   2   3   1M
C1   2   0   0.2533U
.AC LIN 21  5000 15000
.PRINT AC VM(2) VP(2)
.PLOT AC  VM(2) VP(2)
.END
```

```
****      SMALL SIGNAL BIAS SOLUTION        TEMPERATURE =   27.000 DEG C

*******************************************************************************

NODE   VOLTAGE      NODE   VOLTAGE      NODE   VOLTAGE      NODE   VOLTAGE
(   1)   0.0000   (   2)   0.0000   (   3)   0.0000

     VOLTAGE SOURCE CURRENTS
     NAME          CURRENT
     VGEN          0.000E+00
     TOTAL POWER DISSIPATION    0.00E+00   WATTS
```

```
****      AC ANALYSIS                       TEMPERATURE =   27.000 DEG C

*******************************************************************************

   FREQ        VM(2)        VP(2)
   5.000E+03   4.180E-02    8.035E+01
   5.500E+03   4.933E-02    8.008E+01
   6.000E+03   5.850E-02    7.956E+01
   6.500E+03   7.001E-02    7.874E+01
   7.000E+03   8.497E-02    7.752E+01
   7.500E+03   1.053E-01    7.570E+01
   8.000E+03   1.348E-01    7.289E+01
   8.500E+03   1.808E-01    6.826E+01
   9.000E+03   2.608E-01    5.969E+01
   9.500E+03   4.111E-01    4.086E+01
   1.000E+04   5.687E-01   -1.173E+00
   1.050E+04   4.319E-01   -4.320E+01
   1.100E+04   2.882E-01   -6.204E+01
   1.150E+04   2.103E-01   -7.067E+01
   1.200E+04   1.652E-01   -7.539E+01
   1.250E+04   1.364E-01   -7.831E+01
   1.300E+04   1.165E-01   -8.028E+01
   1.350E+04   1.019E-01   -8.170E+01
   1.400E+04   9.079E-02   -8.276E+01
   1.450E+04   8.204E-02   -8.359E+01
   1.500E+04   7.496E-02   -8.425E+01
```

Figure 3.12 Output File ACTANK.OUT.

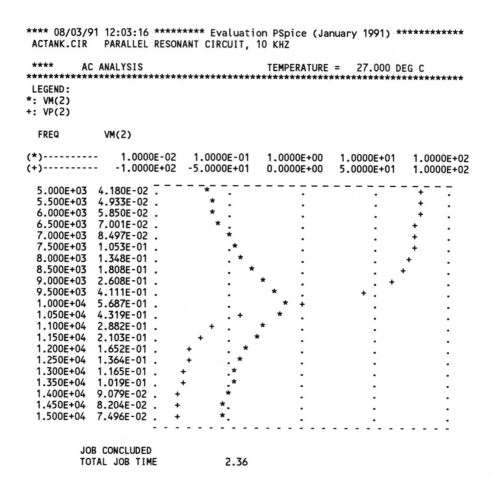

```
   ****     AC ANALYSIS                      TEMPERATURE =   27.000 DEG C
**************************************************************************
  LEGEND:
*: VM(2)
+: VP(2)

   FREQ        VM(2)

 (*)----------   1.0000E-02   1.0000E-01   1.0000E+00   1.0000E+01   1.0000E+02
 (+)----------  -1.0000E+02  -5.0000E+01   0.0000E+00   5.0000E+01   1.0000E+02

   5.000E+03  4.180E-02 .  - - -*- -  .          - - - .        -  .     +  - - -.
   5.500E+03  4.933E-02 .       *  .             .           .       +     .
   6.000E+03  5.850E-02 .       *  .             .           .       +     .
   6.500E+03  7.001E-02 .        * .             .           .      +      .
   7.000E+03  8.497E-02 .         *.             .           .      +      .
   7.500E+03  1.053E-01 .         .*             .           .      +      .
   8.000E+03  1.348E-01 .         . *            .           .     +       .
   8.500E+03  1.808E-01 .         .   *          .           .    +        .
   9.000E+03  2.608E-01 .         .      *       .          .  +           .
   9.500E+03  4.111E-01 .         .         *    .        + .              .
   1.000E+04  5.687E-01 .         .            * +         .               .
   1.050E+04  4.319E-01 .        . +          *  .          .              .
   1.100E+04  2.882E-01 .     +   .         *    .          .              .
   1.150E+04  2.103E-01 .   +     .        *     .          .              .
   1.200E+04  1.652E-01 .  +      .    *         .          .              .
   1.250E+04  1.364E-01 .  +      .   *          .          .              .
   1.300E+04  1.165E-01 . +       . *            .          .              .
   1.350E+04  1.019E-01 . +       . *            .          .              .
   1.400E+04  9.079E-02 . +       .*             .          .              .
   1.450E+04  8.204E-02 . +      *.              .          .              .
   1.500E+04  7.496E-02 . +      *.              .          .              .
                         - - - - .  - - - - - -  .          .      - - - - .
```

```
       JOB CONCLUDED
       TOTAL JOB TIME                   2.36
```

Figure 3.12 (Continued).

output file because of the control lines .PRINT AC VM(2)
VP(2) and .PLOT AC VM(2) VP(2) . The .PRINT command causes a
table to be printed, with columns of frequency, voltage magnitude at node 2, and voltage phase at node 2. The .PLOT command causes a graph of voltage magnitude at node 2 and voltage phase at node 2 to be plotted versus frequency.

The output file, Fig. 3.12, contains the circuit description, a small-signal bias solution (which shows DC values to
be zero, since there are no DC sources in the circuit), a table of results, and a graph of results. The magnitude of the
node 2 voltage is seen to peak at 10 KHz (labeled 1.000E+04
by SPICE), at 568.7 mV. The phase of the voltage at node 2

is 80.35 degrees at 5 KHz, -1.173 degrees at 10 KHz, and -84.25 degrees at 15 KHz; these data show that resonance occurs quite near 10 KHz. Although SPICE can plot up to 8 parameters on one graph, only the first parameter listed in the .PLOT control line will have its data printed as well as plotted on the graph. Plotting more than 3 parameters on the same graph is recommended only for those who love puzzles and have excellent vision. You can use as many .PLOT control lines as needed, so it is never necessary to put more than 3 plots on one graph.

3.7 THE FOURTH CIRCUIT - RC CIRCUIT WITH PULSE INPUT

In this example, an RC circuit will be connected to a pulse voltage source, and an analysis of circuit voltage versus time, called a transient analysis, will be performed. The result of the transient analysis will appear in the output file in tabular form and as a graph of voltage versus time. The circuit is shown in Fig. 3.13. The single pulse input source goes from 0 V to 1 V after 10 microseconds of delay (relative to the beginning of the transient analysis), with rise and fall times of 1 nanosecond, and has a duration of 80 microseconds. It is described in the SPICE input file by the element line "VIN 1 0 PULSE(0 1 10U 1N 1N 80U)". The results of the

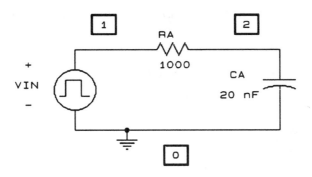

Figure 3.13 R-C Circuit with Pulse Input.

```
PULSE.CIR    R-C CIRCUIT WITH PULSE INPUT
VIN 1 0 PULSE(0 1 10U 1N 1N 80U)
RA  1  2  1000
CA  2  0  20N
.TRAN 5U 160U
.PRINT TRAN V(2) V(1)
.PLOT  TRAN V(2) V(1)
.END
```

Figure 3.14 Input File PULSE.CIR.

transient analysis will appear in the output file because of
the control lines .PRINT TRAN V(2) V(1) and .PLOT TRAN V(2)
V(1). The input file is shown in Fig. 3.14.

Since there are no DC sources in the circuit, and the
pulse level is 0 V at time zero, the output file (Fig. 3.15)
indicates that at time zero the voltage at nodes 1 and 2 is 0
V. The table of transient analysis results and the graph of
V(1) and V(2) versus time show that the input voltage, V(1),
is 0 V up through 10 usec, is 1 V from 15 to 90 usec, and returns to 0 V at 95us where it remains until 160 usec. The
voltage across the capacitor, V(2), begins rising exponentially at 10 usec, reaches 981.8 mV at 90 usec, and exponentially
decays to 29.44 mV at 160 usec. The input voltage and the capacitor voltage are identical or nearly so at 0, 5, 10 and 90
usec. This overlap of the two graphs is indicated by an X
symbol at those four times.

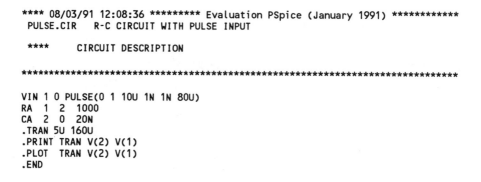

```
**** 08/03/91 12:08:36 ********* Evaluation PSpice (January 1991) ************
PULSE.CIR    R-C CIRCUIT WITH PULSE INPUT

****        CIRCUIT DESCRIPTION

*****************************************************************************

VIN 1 0 PULSE(0 1 10U 1N 1N 80U)
RA  1  2  1000
CA  2  0  20N
.TRAN 5U 160U
.PRINT TRAN V(2) V(1)
.PLOT  TRAN V(2) V(1)
.END
```

Figure 3.15 Output File PULSE.OUT.

```
**** 08/03/91 12:08:36 ********* Evaluation PSpice (January 1991) ************
PULSE.CIR   R-C CIRCUIT WITH PULSE INPUT

****       INITIAL TRANSIENT SOLUTION          TEMPERATURE =   27.000 DEG C


*********************************************************************************

NODE    VOLTAGE     NODE   VOLTAGE     NODE   VOLTAGE     NODE   VOLTAGE
(   1)   0.0000  (   2)    0.0000

    VOLTAGE SOURCE CURRENTS
    NAME         CURRENT
    VIN          0.000E+00

    TOTAL POWER DISSIPATION   0.00E+00  WATTS


**** 08/03/91 12:08:36 ********* Evaluation PSpice (January 1991) ************
PULSE.CIR   R-C CIRCUIT WITH PULSE INPUT

****       TRANSIENT ANALYSIS               TEMPERATURE =   27.000 DEG C


*********************************************************************************

    TIME        V(2)        V(1)
    0.000E+00   0.000E+00   0.000E+00
    5.000E-06   0.000E+00   0.000E+00
    1.000E-05   2.103E-08   6.400E-06
    1.500E-05   2.189E-01   1.000E+00
    2.000E-05   3.932E-01   1.000E+00
    2.500E-05   5.271E-01   1.000E+00
    3.000E-05   6.321E-01   1.000E+00
    3.500E-05   7.138E-01   1.000E+00
    4.000E-05   7.771E-01   1.000E+00
    4.500E-05   8.269E-01   1.000E+00
    5.000E-05   8.650E-01   1.000E+00
    5.500E-05   8.952E-01   1.000E+00
    6.000E-05   9.182E-01   1.000E+00
    6.500E-05   9.364E-01   1.000E+00
    7.000E-05   9.505E-01   1.000E+00
    7.500E-05   9.615E-01   1.000E+00
    8.000E-05   9.701E-01   1.000E+00
    8.500E-05   9.767E-01   1.000E+00
    9.000E-05   9.819E-01   1.000E+00
    9.500E-05   7.652E-01   0.000E+00
    1.000E-04   5.970E-01   0.000E+00
    1.050E-04   4.633E-01   0.000E+00
    1.100E-04   3.615E-01   0.000E+00
    1.150E-04   2.810E-01   0.000E+00
    1.200E-04   2.188E-01   0.000E+00
    1.250E-04   1.703E-01   0.000E+00
    1.300E-04   1.324E-01   0.000E+00
    1.350E-04   1.032E-01   0.000E+00
    1.400E-04   8.008E-02   0.000E+00
    1.450E-04   6.251E-02   0.000E+00
    1.500E-04   4.855E-02   0.000E+00
    1.550E-04   3.784E-02   0.000E+00
    1.600E-04   2.939E-02   0.000E+00
```

Figure 3.15 (Continued).

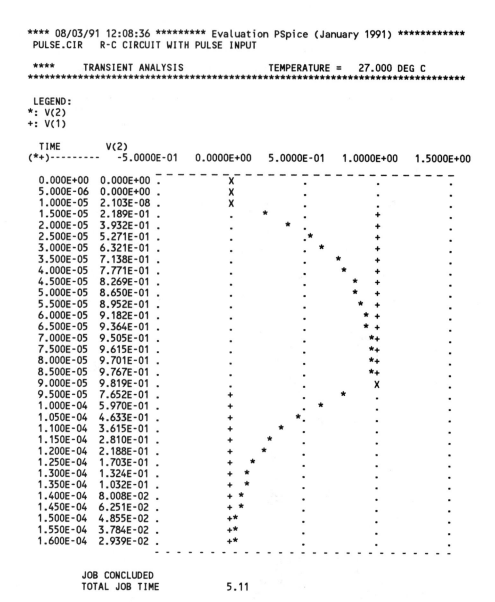

```
****      TRANSIENT ANALYSIS                TEMPERATURE =   27.000 DEG C
*****************************************************************************

 LEGEND:
 *: V(2)
 +: V(1)

  TIME          V(2)
 (*+)---------  -5.0000E-01   0.0000E+00   5.0000E-01   1.0000E+00   1.5000E+00

 0.000E+00  0.000E+00 .            X              .            .            .
 5.000E-06  0.000E+00 .            X              .            .            .
 1.000E-05  2.103E-08 .            X              .            .            .
 1.500E-05  2.189E-01 .            .       *      .            +            .
 2.000E-05  3.932E-01 .            .          *   .            +            .
 2.500E-05  5.271E-01 .            .            .*             +            .
 3.000E-05  6.321E-01 .            .            .   *          +            .
 3.500E-05  7.138E-01 .            .            .      *       +            .
 4.000E-05  7.771E-01 .            .            .        *     +            .
 4.500E-05  8.269E-01 .            .            .          *   +            .
 5.000E-05  8.650E-01 .            .            .           *  +            .
 5.500E-05  8.952E-01 .            .            .            * +            .
 6.000E-05  9.182E-01 .            .            .            *  +           .
 6.500E-05  9.364E-01 .            .            .            *  +           .
 7.000E-05  9.505E-01 .            .            .            .  *+          .
 7.500E-05  9.615E-01 .            .            .            .  *+          .
 8.000E-05  9.701E-01 .            .            .            .  *+          .
 8.500E-05  9.767E-01 .            .            .            .  *+          .
 9.000E-05  9.819E-01 .            .            .            .   X          .
 9.500E-05  7.652E-01 .            +            .         *  .              .
 1.000E-04  5.970E-01 .            +            .    *       .              .
 1.050E-04  4.633E-01 .            +          *  .           .              .
 1.100E-04  3.615E-01 .            +      *      .           .              .
 1.150E-04  2.810E-01 .            +    *        .           .              .
 1.200E-04  2.188E-01 .            +  *          .           .              .
 1.250E-04  1.703E-01 .            + *           .           .              .
 1.300E-04  1.324E-01 .            + *           .           .              .
 1.350E-04  1.032E-01 .            + *           .           .              .
 1.400E-04  8.008E-02 .            + *           .           .              .
 1.450E-04  6.251E-02 .            + *           .           .              .
 1.500E-04  4.855E-02 .            +*            .           .              .
 1.550E-04  3.784E-02 .            +*            .           .              .
 1.600E-04  2.939E-02 .            +*            .           .              .

           JOB CONCLUDED
           TOTAL JOB TIME          5.11
```

Figure 3.15 (Continued).

CHAPTER SUMMARY

Four examples of using PSpice for circuit analysis have been presented in this chapter:

    1. A DC circuit with one battery, without any control lines in the input file. PSpice by default performed a

"Small Signal Bias Solution" which gave the node voltages. A slight change to the input file showed the paper-saving effects of adding a .OPTIONS NOPAGE control line. See input files 3R.CIR and 3RNOPAGE.CIR.

2. A DC circuit with two batteries; one battery was stepped through a range of voltages to illustrate performing a DC analysis. The results were plotted and presented in tabular form. See input file 3R2BAT.CIR.

3. A parallel resonant (tank) circuit was stepped over a range of frequencies, and an AC analysis was performed at each frequency for the voltage across the tank. The results were presented in tabular form and plotted. See input file ACTANK.CIR.

4. A series resistor-capacitor was connected to an input voltage pulse. A transient analysis was done, and the exponentially varying capacitor voltage was shown in tabular form and plotted. See input file PULSE.CIR.

By looking at these examples, and modifying them as suggested in the problems section which follows, you will gain proficiency at "talking" in PSpice language and using it for problems of your own.

PROBLEMS

1. Refer to the circuit schematic diagram in Fig. 3.16, which has nodes notated for PSpice analysis.
    a. Write a PSpice input file describing the circuit, including a title line and a .END line.
    b. Analyze the circuit for the DC voltage at node 12 (com-

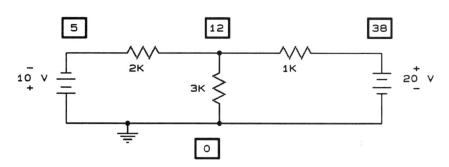

Figure 3.16 Circuit for Prob. 1.

pared to ground), using conventional circuit analysis methods.

c. **Run a PSpice analysis using the input file from Prob. 1.a.** How does your answer in 1.b compare with the PSpice analysis result?

2. Refer to the circuit diagram in Fig. 3.10.
   a. Create an input file for that circuit, but replace the 3 ohm resistor with a 1 ohm resistor.
   b. Predict what will happen to the shape of the graph of voltage magnitude at node 2 versus frequency. (Hint: a smaller resistor will affect the circuit Q, or quality factor).
   c. Run a PSpice analysis using the input file from 2.a. Compare the graph that results with your prediction from 2.b.

3. In Fig. 3.13, replace the 20 nF capacitor with a 20 mH inductor, and repeat the PSpice analysis.

4. Use PSpice to find the voltage at node 16 in Fig. 3.17.

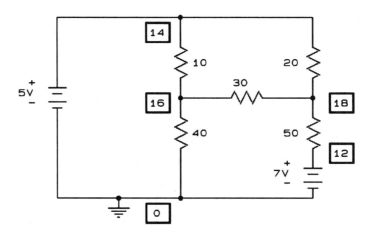

Figure 3.17 Circuit for Prob. 4.

Chapter 4

LINEAR DEPENDENT SOURCES

4.1 INTRODUCTION

PSpice allows the user to use dependent (or controlled) voltage sources and current sources in describing circuits. This can be useful in making simple models of semiconductor devices such as bipolar junction transistors, field-effect transistors and operational amplifiers. In this chapter linear controlled sources will be introduced; Appendix C presents the topic of nonlinear dependent sources.

The four kinds of linear dependent sources are:

voltage-controlled current source, or VCCS
voltage-controlled voltage source, or VCVS
current-controlled current source, or CCCS
current-controlled voltage source, or CCVS

The sources are uniquely specified by an element name beginning with the letter G, E, F or H, and have units as described below:

ELEMENT	EQUATION	ELEMENT TYPE	UNIT
VCCS	$I = G (V)$	Transconductance	Siemen
VCVS	$V = E (V)$	Voltage gain	V/V
CCCS	$I = F (I)$	Current gain	A/A
CCVS	$V = H (I)$	Transresistance	Ohm

4.2 LINEAR VCCS

Linear voltage-controlled current sources are described by the following element line

GXXXXXX   N+ N- NC+ NC- VALUE

where N+ and N- are the nodes to which the current source is connected, NC+ and NC- are the nodes which define the controlling voltage, and VALUE is the transconductance of the VCCS in siemens.

EXAMPLE:

GWHIZ 4 13 22 26 5E-2
    The current source connected between nodes 4 and 13 (with positive current assumed to flow from 4 to 13) is controlled by the voltage difference between nodes 22 and 26 (node 22 considered positive).   The transconductance is 50 millisiemens. See Fig. 4.1.

In order to illustrate the use of the VCCS element, a simple model of a JFET will be done with a VCCS. The circuit in Fig. 4.2, with nodes notated for PSpice analysis, can be redrawn with a VCCS replacing the JFET.   See Fig. 4.3.   The element lines describing the redrawn circuit are shown below.   The transconductance of the JFET and the VCCS is assumed to be 20 mS.

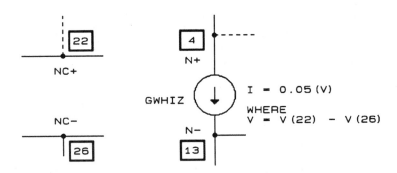

Figure 4.1 Voltage-Controlled Current Source.

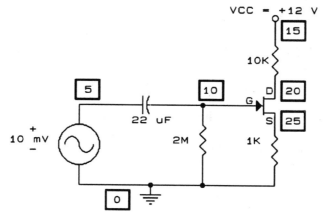

Figure 4.2 JFET Circuit Notated for PSpice.

VCC	15	0	12	
VSIGNAL	5	0	AC	10M
CGATE	5	10	22U	
RGATE	10	0	2MEG	
RSOURCE	25	0	1K	
RDRAIN	15	20	10K	
GFET	20	25	10	25	20E-3

In this JFET modeling example, the current that flows from drain to source is determined by the gate-source voltage, V_{GS}, and the JFET transconductance of 20 mS:

$$I_{DS} = .020 \, (V_{GS})$$

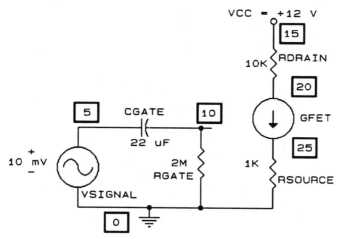

Figure 4.3 JFET Circuit Modeled with VCCS.

4.3 LINEAR VCVS

Linear voltage-controlled voltage sources are described by the following element line

EXXXXXX   N+ N- NC+ NC- VALUE

where N+ and N- are the nodes to which the voltage source is connected,   NC+ and NC- are the nodes which define the controlling voltage, and VALUE is the voltage gain of the VCVS in V/V.

EXAMPLE:

EGAIN 14 9 6 2 250

The voltage source connected between nodes 14 and 9 (node 14 is positive) has a magnitude of 250 times the voltage between nodes 6 and 2. Refer to Fig. 4.4.

A good example of the use of the VCVS element is as a small-signal model of an operational amplifier at low frequencies (where slew-rate and gain-bandwidth limitations can be neglected).   The circuit in Fig. 4.5, with nodes notated for PSpice analysis, can be redrawn with an input resistance and a VCVS replacing the op-amp.   The operational amplifier has a differential input resistance of 1 megohm and an open-loop gain of 50,000 V/V.   See Fig. 4.6.   The element lines describing the redrawn circuit are shown below.

VSIG	4	0	AC 1
RSIG	4	11	50
RFB	22	19	20K
R1	19	0	1K

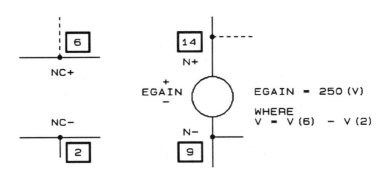

Figure 4.4   Voltage-Controlled Voltage Source.

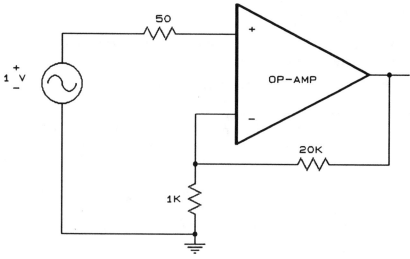

Figure 4.5 Op-Amp Circuit.

```
RINOPAMP    11   19   1MEG
EOPAMP      22   0    11   19   5E4
```

In this op-amp model, the voltage at the output of the op-amp
is equal to the input differential voltage (voltage at node 11
minus the voltage at node 19) multiplied by the open-loop gain
of 50,000 V/V.   Of course, this simple model neglects saturation, capacitance and slew-rate, among others.

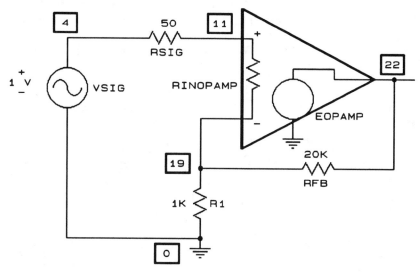

**Figure 4.6 Op-Amp Circuit Modeled with
VCVS.**

4.4 LINEAR CCCS

Linear current-controlled current sources are described by the following element line

  FXXXXXX   N+ N- VNAME VALUE

where N+ and N- are the nodes to which the current source is connected, VNAME is the name of a voltage source in the same input file through which the current controlling the CCCS flows, and VALUE is the current gain of the CCCS in A/A. Note that it may be necessary to insert a dead voltage source in the circuit branch where the controlling current flows in order to have a VNAME to control the CCCS. Such a dead voltage source would have no effect on circuit operation.

EXAMPLE:

  FOUT 37 24 VSENSE 5

    Refer to Fig. 4.7. The current in CCCS FOUT connected between nodes 37 and 24 has a magnitude of 5 times the current flowing (downward from node 6 to node 11) through voltage source VSENSE and the 20 ohm resistor. Positive current in FOUT flows from node 37 to node 24, through the current source.

Bipolar junction transistors can be thought of as current-controlled current sources for small-signal operation. The BJT circuit shown in Fig. 4.8 will be modeled as a resistance for the base-emitter junction and as a CCCS for the collector-emitter junction. The forward current gain, beta, is assumed

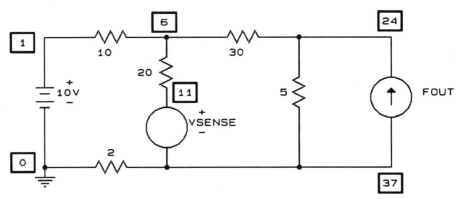

Figure 4.7 Current-Controlled Current Source Circuit.

Linear Dependent Sources   *Chap. 4*

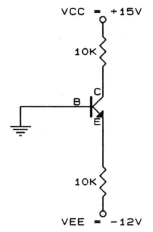

Figure 4.8 Bipolar Junction Transistor Circuit.

to be 80 A/A. Figure 4.9 shows the modeled circuit. The input element lines for Fig. 4.9 are shown below.

VIB	0	2	0
FBJT	4	5	VIB 80
VCC	3	0	15
VEE	6	0	-12
RBASE	2	5	1.8K
RE	5	6	10K
RC	3	4	10K

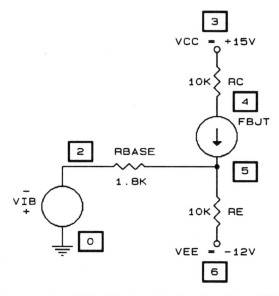

Figure 4.9 BJT Circuit Modeled with CCCS.

The resistance between node 2 and 5 (RBASE) is the hybrid parameter (hie), for which 1.8K ohms is appropriate for this circuit. Note that the dummy voltage source VIB (its magnitude is zero) was added to the circuit in order to provide a way of measuring the base current flowing up from ground into the base terminal of the model (node 2). VIB's positive node is 0 and its negative node is 2; this way current flowing from node 0 to node 2 is considered positive. While this is not a very practical model of a BJT, it does illustrate the format for a CCCS element line.

4.5 LINEAR CCVS

Linear current-controlled voltage sources are described by the following element line

> HXXXXXX   N+ N- VNAME VALUE

where N+ and N- are the nodes to which the voltage source is connected, VNAME is the name of a voltage source in the same input file through which the current controlling the CCVS flows, and VALUE is the transresistance of the CCVS in ohms. Note that it may be necessary to insert a dead voltage source in the circuit branch where the controlling current flows, in order to have a VNAME to control the CCVS. Such a dead voltage source would have no effect on circuit operation.

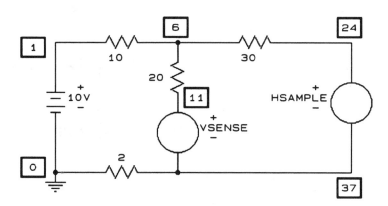

Figure 4.10 Current-Controlled Voltage Source Circuit.

EXAMPLE:

HSAMPLE   24 37 VSENSE 6

Refer to Fig. 4.10. The voltage across CCVS HSAMPLE, connected between nodes 24 and 37, has a magnitude of 6 times the current flowing (downward from node 6 to node 11) through voltage source VSENSE and the 20 ohm resistor.

This chapter has been devoted to presenting linear dependent sources. PSpice is capable of analyzing circuits containing non-linear dependent sources as well. Non-linear dependent sources are discussed in Appendix C.

CHAPTER SUMMARY

PSpice has four kinds of linear dependent sources: VCCS, VCVS, CCCS and CCVS.

Linear dependent sources can be used to create your own models of circuit devices.

All four kinds of linear dependent sources are connected to two nodes. Two kinds (VCCS and VCVS) are controlled by the difference in voltage between two nodes, while the other two kinds (CCCS and CCVS) are controlled by the current through an independent voltage source. This independent voltage source (which may be a dead source) may have to be added to a branch to sense the current in that branch.

Chapter 5

TYPES OF ANALYSIS

5.1 INTRODUCTION

By including control lines in an input file, PSpice can be made to perform many kinds of analyses of a circuit. Generally, only one or two will be done at a time. Later in this chapter, use of each control line will be illustrated with examples. The many examples in Chap. 12, including circuit schematic diagrams, input files, analysis results and a discussion of each example, will be helpful as well. The control lines and a brief description of the types of analysis are:

.DC   A non-linear analysis which determines the DC operating point of the circuit. Capacitors are open-circuited, and inductors are short-circuited during DC analysis. One or more sources (current and voltage) may be stepped over a range. Results of the analysis at each source value may be printed in a table and/or plotted.

.AC   This is a linear small-signal analysis. PSpice determines the DC operating point of a circuit and thereby calculates values for small-signal models of all nonlinear devices (semiconductors, inductors, capacitors, polynomial dependent sources). Then the linear small-signal circuit model is analyzed at each frequency specified by the user.

Results of the analysis at each frequency can be printed and/or plotted.

.TRAN   This transient analysis determines the output variables (current and voltage) as a function of time over a time interval which is specified.   The size of the time step used within the interval may be specified.   A table can be printed and/or a plot made of the output variables versus time.

.OP   Causes PSpice to solve for and print the DC operating point of a circuit.   This is done automatically when an AC analysis is performed, to find the AC small-signal model parameters, and also when a transient analysis is done, to determine initial conditions.

.TF   Does a small-signal DC transfer function analysis of a circuit from a specified input to a specified output. The input resistance, output resistance, and transfer function (voltage gain, current gain, transresistance or transconductance) are printed.

.SENS   PSpice finds the DC small-signal sensitivities of one or more specified output variables with respect to every parameter in the circuit, including semiconductor parameters.   A large amount of paper can be consumed printing the results of this analysis for any but the smallest circuits !

.NOISE   When present with a .AC control line, PSpice will find the equivalent output noise and equivalent input noise at specified output and input points in the circuit. Results may be printed and/or plotted.

.FOUR   Performs a Fourier analysis of an output variable when done with a transient analysis.   Computes the amplitudes and phases of the first nine frequency components (harmonics) of a user-specified fundamental frequency, as well as the DC component.

.TEMP   Not truly an analysis by itself, .TEMP tells PSpice at what temperature(s) to simulate a circuit.

5.2   DC ANALYSIS USING .DC CONTROL LINE

Control line format:

        .DC SRC START STOP INCR <SRC2 START2 STOP2 INCR2>

In the control line above, SRC is the name of the independent voltage or current source being varied.   START is the starting value, STOP is the final value, and INCR is the increment size; units are volts or amps. Optionally, a second source (SRC2) may be specified, along with its associated parameters. If it is, the first source will be swept over its entire range (START to STOP) for each value of the second source. Associated with any .DC control line should be a .PRINT or a .PLOT control line which will cause an output to be written to the PSpice output file. For more information on .PRINT and .PLOT, see Chap. 6.

EXAMPLES:

.DC VGATE -2 3 0.5

Independent voltage source VGATE (specified elsewhere in the input file) is incremented from -2 V to +3 V in 0.5 V steps. In other words, it will be set to values of -2, -1.5, -1, . . . , +3 V.

.DC IBASE 0 100U 10U VCE 0 20 4

Independent current source IBASE will step from 0 to 100 uA in steps of 10 uA while independent voltage source VCE is fixed at 0 V. Then VCE will jump to 4 V, and IBASE will step again from 0 to 100 uA. This process will be repeated until VCE reaches 20 V. A total of 66 DC analyses will occur; 11 for IBASE, multiplied by 6 for VCE.

Alternate formats for the DC analysis control line are:

        .DC DEC SRC START STOP ND <SRC2 START2 STOP2 ND>
        .DC OCT SRC START STOP NO <SRC2 START2 STOP2 NO>
        .DC SRC LIST VALUE1 VALUE2 . . .
Note: For detailed information on alternate formats for DC analysis control lines, refer to the PSpice User's Guide, available from MicroSim Corporation.

5.3 AC ANALYSIS USING .AC CONTROL LINE

Control line format:

        .AC LIN NP FSTART FSTOP
        .AC DEC ND FSTART FSTOP
        .AC OCT NO FSTART FSTOP

In the control lines above, FSTART is the lowest frequency (which may not be negative or zero), and FSTOP is the highest frequency. The frequencies at which the analyses are done can be specified three ways:

LIN: NP will cause an analysis to occur at a number of frequencies equal to NP, linearly spaced between FSTART and FSTOP. To do an AC analysis at a single frequency, set NP equal to 1, and make FSTART and FSTOP equal the desired frequency.

DEC: ND will make PSpice divide the frequency range into decades, with ND frequencies per decade. Frequencies will be logarithmically-spaced (the ratio of any two adjacent frequencies is the same over the entire range). This decade spacing is most useful with wide ranges of frequency; if FSTOP is not an integer number of decades above FSTART, the highest frequency can exceed FSTOP.

OCT: NO instructs PSpice to break the frequency range into octaves, with NO frequencies per octave. Frequencies will be logarithmically-spaced (the ratio of any two adjacent frequencies is the same over the entire range). This octave spacing is most useful with wide ranges of frequency; if FSTOP is not an integer number of octaves above FSTART, the highest frequency can exceed FSTOP.

Unlike the .DC control line, the .AC control line does not specify which source(s) is to be swept in frequency. The frequency of all AC sources in the circuit will be set simultaneously by the .AC control line. Associated with any .AC control line should be a .PRINT or a .PLOT control line which will cause an output to be written to the PSpice output file. For more information on .PRINT and .PLOT, see Chap. 6.

EXAMPLES:

.AC LIN 101 5000 6000

Any independent AC sources in the circuit will start at 5 KHz and increment upward by 10 Hz until 6 KHz is reached. Thus, AC analysis will occur at frequencies of 5000, 5010, 5020, . . . , 5990, 6000 Hz.

.AC LIN 1 3.58MEG 3.58MEG

All independent AC sources are set to 3.58 MHz for the AC analysis.

.AC DEC 5 1 10K

Since there are 4 decades of frequency [log(10K/1) = 4], and 5 points per decade, AC analysis will occur at 21 frequencies logarithmically-spaced between 1 Hz and 10 KHz.

.AC OCT 7 10 5120

The number of octaves of frequency can be determined by: NO = [(log(5120/10))/log2] = 9. AC analysis will be done at 64 frequencies (7 points/octave, 9 octaves) spaced logarithmically between 10 Hz and 5.12 KHz.

5.4 TRANSIENT ANALYSIS USING .TRAN CONTROL LINE

Control line format:

.TRAN TSTEP TSTOP <TSTART <TMAX>> <UIC>

In the control line above, TSTEP is the time increment used for plotting and/or printing results of the transient analysis; it is not necessarily the computing time step that PSpice uses between successive analyses. TSTOP is the time of the last transient analysis. TSTART may be omitted; its default value is zero. PSpice always starts the analysis at time zero. If a TSTART value greater than zero is specified, PSpice will analyze the circuit from zero to TSTART but will not store the analysis results.

The largest computing time step PSpice will use is TMAX. If TMAX is not specified, the computing time step will be the smaller of TSTEP or (TSTOP - TSTART)/50. TMAX can be used to ensure that the computing time step is smaller than the plotting/printing interval, TSTEP.

The parameter UIC means use initial conditions. When UIC appears in the .TRAN control line, PSpice does not compute the quiescent operating point of the circuit prior to doing the transient analysis. PSpice will do one of two things:

1. If a .IC control line is not present, it will use the initial conditions specified on element lines (capacitors, inductors and semiconductors) as the starting point for its analysis.

2. If a .IC control line is present, it will use the node

voltages on the .IC line to determine the initial conditions for elements.

EXAMPLES:

.TRAN 1U 80U

A transient analysis will occur every 1 us from time zero to 80 us. 81 points will be plotted or printed.

.TRAN 2N 1000N

The intent of this control line is to do a transient analysis every 2 nanoseconds, up to 1000 nanoseconds. However, an error will occur here because the number of points to be plotted or printed is 501, in excess of the PSpice default limit of 201. This problem is easily corrected by using the "LIMPTS =" command in a .OPTIONS control line. For example, the control line .OPTIONS LIMPTS = 600 would prevent an error message. See Appendix A.

.TRAN 5M 500M 100M 2M UIC

Every 5 ms from time zero up to 500 ms a transient analysis will be done; however, results will not be stored (or plotted or printed) until 100 ms. The computing time step will be 2 ms (TSTEP is 5 ms, [TSTOP - TSTART]/50 = [500 ms - 100 ms] /50 = 8 ms). If 2M had not appeared in the .TRAN control line, the computing time step would have been the smaller of 5 ms and 8 ms, or 5 ms. Initial conditions will be used for the first transient analysis.

5.5 OPERATING POINT ANALYSIS USING .OP CONTROL LINE

Control line format:

.OP

That's it, plain and simple. There are no optional parameters one can specify. This control line instructs PSpice to solve for the operating point of the circuit with capacitors opened and inductors shorted, and to print detailed results of the operating point analysis. Such things as total circuit power dissipation and currents of voltage sources, as well as small-signal parameters of semiconductors and nonlinear devices, are printed.

PSpice performs a DC operating point analysis without the .OP control line when an AC analysis or transient analysis is performed, but it does not print such detailed results.

5.6 TRANSFER FUNCTION ANALYSIS USING .TF CONTROL LINE

Control line format:

.TF OUTPUTVAR INPUTSRC

OUTPUTVAR is a small-signal output variable (voltage or current). INPUTSRC is a small-signal input source (voltage or current). The .TF control line does a small-signal DC transfer function analysis, and will print the input resistance at INPUTSRC, the output resistance at OUTPUTVAR, and some type of transfer function (voltage gain, current gain, transresistance or transconductance) in the output file.

EXAMPLES:

.TF V(5) VIN

The specified output is the node 5 voltage, and the specified input is a voltage source already defined in the input file as VIN. The transfer function would be a voltage gain.

.TF I(VIDRAIN) VGATE

The output is the current through voltage source VIDRAIN (likely to be a dummy voltage source used as an ammeter in the drain branch of a FET), while the input is a voltage source already defined in the input file as VGATE. The transfer function would be current/voltage, or transconductance.

.TF V(12,15) IBASE

The output is the difference in voltage between nodes 12 and 15 (perhaps collectors in a BJT differential amplifier), and the input is a current source previously defined in the input file as IBASE. The transfer function is the quotient of voltage/current, or transresistance.

5.7 SENSITIVITY ANALYSIS
USING .SENS CONTROL LINE

Control line format:

    .SENS OV1 <OV2> ...

OV1 is the output variable such as node voltage or current through a voltage source. One output variable must be specified; others are optional. The DC small-signal sensitivity of each output variable to each circuit parameter will be determined. It is easy to fell entire forests due to massive amounts of paper consumed when this control line is used in any but the simplest circuits. A PSpice analysis of a simple two-transistor differential amplifier with 3 resistors, 2 BJTs and two batteries printed 34 lines of sensitivity information with only one output variable specified. Each BJT contributed 14 lines to the list.

EXAMPLES:

.SENS V(5) I(VSUPPLY)

The small-signal DC sensitivity of the voltage at node 5 will be determined for each parameter in the circuit. In addition, the small-signal DC sensitivity of the current in voltage source VSUPPLY (defined elsewhere in the input file) will be determined for each parameter in the circuit.

5.8 NOISE ANALYSIS USING
.NOISE CONTROL LINE

Control line format:

    .NOISE OUTPUTV INPUTSRC NUMSUM

This control line must be used with a .AC control line. PSpice will perform a noise analysis of the circuit along with the AC small-signal steady-state analysis. OUTPUTV is a voltage which will be considered the output summing point for noise. INPUTSRC is the element name of an independent voltage or current source which will be the noise input reference. NUMSUM is the summary interval; it is the interval at which a summary printout of the contributions of all noise generators (resistors, semiconductors) is printed. If NUMSUM is omitted or set to zero, no summary printout is done. For example, if NUMSUM is set to 5, then a summary printout will occur at every fifth AC analysis frequency.

This control line too generates many lines of information in the output file if NUMSUM is set to 1. Results of the noise analysis can also be plotted and/or printed. See Chap. 6 for information on doing this.

EXAMPLES:

.NOISE V(2,3) VIN1 10

A noise analysis will be done along with the AC analysis. The summary printout will list, at every tenth frequency, the output noise measured as the difference between nodes 2 and 3, and the equivalent input noise referenced to voltage source VIN1.

.NOISE V(17) IBIAS

A noise analysis is done with the output noise measured at node 17, and the equivalent input noise referenced to current source IBIAS. Although no summary printouts will occur, the output noise and equivalent input noise can be plotted and/or printed at each frequency of the AC analysis.

5.9 FOURIER ANALYSIS USING .FOUR CONTROL LINE

Control line format:

.FOUR FREQ OV1 <OV2 OV3 ...>

A Fourier analysis can be done only in conjunction with a transient analysis (a .TRAN control line must appear in the same input file). The .FOUR control line instructs PSpice to determine the amplitudes of the DC component and the first nine frequency components of a waveform (the fundamental and second through eighth harmonics). Results are written to the output file without the need for .PLOT or .PRINT control lines.

FREQ is the fundamental frequency (determined by you). If the input to a circuit was a triangle waveform with a period of 5 ms, the fundamental frequency would be 1/(5 ms), or 200 Hz. OV1 . . . are the output variables on which a Fourier analysis is to be performed. It is important to note that the Fourier analysis is not performed over the entire time of the transient analysis. It is done only at the very end of the time, from (TSTOP-period) to TSTOP. Thus one must be sure to make the transient analysis at least 1/FREQ long. For example, for the .FOUR analysis to be meaningful, a circuit with a a 60 Hz sinusoid input should be at least 1/(60 Hz) or 16.7 ms long.

In order to achieve good accuracy in the Fourier analysis results, the maximum computing time step (TMAX in the .TRAN control line) should be set to period/100 or less.

EXAMPLES:

.FOUR 60 V(6) I(VSOURCE)

A Fourier analysis will be done on the voltage at node 6 and also on the current through VSOURCE. The user-specified fundamental frequency is 60 Hz.

.FOUR 2MEG V(8,2)

Using a fundamental frequency of 2 MHz, a Fourier analysis will be done on the voltage between nodes 8 and 2.

5.10   ANALYSIS AT DIFFERENT TEMPERATURES USING .TEMP CONTROL LINE

Control line format:

.TEMP T1 <T2 <T3 . . . >>

The .TEMP line sets the temperature(s) at which PSpice will simulate the circuit. T1, T2, . . . are the temperatures expressed in centigrade. If more temperatures than T1 are specified, all analyses are performed at each temperature. Temperatures less than -223.0 C will be ignored by PSpice.

PSpice will assume that the nominal circuit temperature, TNOM, is 27 C unless the TNOM option in the .OPTIONS control line is used to set another value. Model parameters for semiconductors are assumed to be valid at TNOM. TNOM also affects the resistance variation of temperature-dependent resistors.

EXAMPLES:

.TEMP 0 25 60 100

All analyses specified in the circuit input file will be performed at 0, 25, 60 and 100 degrees centigrade.

.TEMP -40

All analyses specified in the circuit input file will be performed at -40 degrees centigrade.

CHAPTER SUMMARY

PSpice analyses include DC, AC, transient, operating point, transfer function, sensitivity, noise and Fourier. In addition, the circuit temperature can be set to any value(s) with the .TEMP control line.

AC analysis can be done over a range of many frequencies which are linearly- or logarithmically-spaced.

Some of the analyses can create <u>very</u> large output files; look at the output file on the monitor before printing it to save paper.

PROBLEMS

1. Refer to the circuit schematic diagram in Fig. 5.1, which has nodes notated for PSpice analysis.
    a. Write a PSpice input file describing the circuit, including a title line and a .END line.

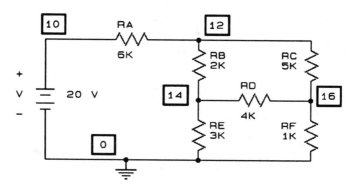

Figure 5.1 Circuit for Prob. 1.

b. Analyze the circuit for the DC voltage across the 4K ohm resistor (including polarity), using conventional circuit analysis methods.

c. Run a PSpice analysis using the input file from Prob. 1.a. How does your answer in 1.b compare with the results obtained from the PSpice analysis?

2. The circuit of Fig. 5.2 contains a voltage-controlled voltage source, E. The voltage across E, V(4,0), is given by

$$V(4,0) = 7*VA$$

where VA is the voltage across the 1 ohm resistor, V(2,1).

a. Write a PSpice input file describing the circuit.

b. Use PSpice to solve for the current through the 2 ohm resistor (a dead voltage source is included in the circuit to serve as an ammeter).

c. Modify the input file to determine the sensitivity of the current through the 2 ohm resistor to the magnitude of the 3 V battery. Run the analysis.

3. A filter circuit used in frequency modulators is shown in Fig. 5.3. It is called a pre-emphasis circuit, and has an interesting response as a function of frequency. The response is flat up to about 2 KHz, then rises at 20 dB/decade until about 20 KHz, above which the response is flat again. This filter increases the amplitude of the frequencies above 2 KHz prior to modulation. In the receiver, after demodulation a similar filter with the opposite response attenuates frequen-

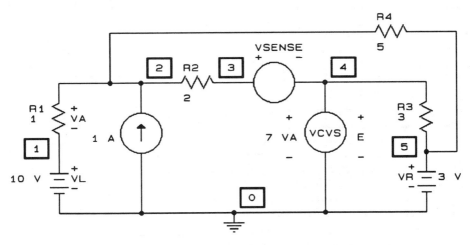

Figure 5.2 Circuit for Prob. 2.

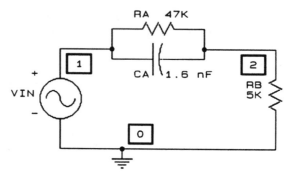

Figure 5.3 Circuit for Prob. 3.

cies (and noise) above 2 KHz, improving the signal to noise
ratio at the output.

    a. Use PSpice to do an AC analysis from 200 Hz to 200 KHz,
with logarithmically-spaced frequencies. Plot the output
voltage magnitude at node 2 versus frequency.

4. Figure 5.4 is alleged to be a notch filter. It consists
of a 2-pole high-pass filter in cascade with a 2-pole low-
pass filter, connected to the inverting input of a true differential amplifier. The input signal feeds both the high-
pass filter and the non-inverting input of the differential

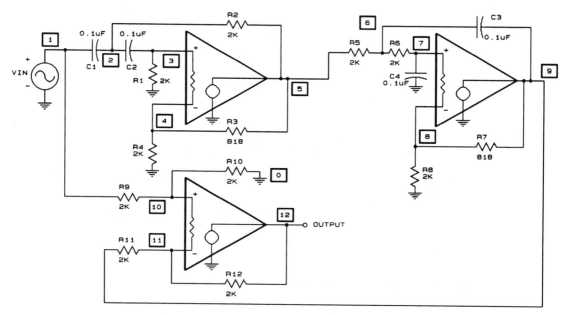

Figure 5.4 Circuit for Prob. 4.

amplifier. Each op-amp is modeled as an input resistance of 1E6 ohms, and a VCVS with a gain of 1E5 V/V.

a. Write a PSpice input file describing the circuit. Include an AC analysis which sweeps the frequency of VIN, and a plot of the gain versus frequency. For, convenience, make VIN have a magnitude of 1 V. Choose a minimum and maximum frequency, as well as a type of sweep (LIN, DEC and OCT are available) which will show the notch frequency in a graph of output (node 12) voltage versus frequency.

b. Based on the result of 4.a, narrow the sweep range so that the shape and depth of the notch are shown clearly. What is the gain of the circuit at the center of the notch?

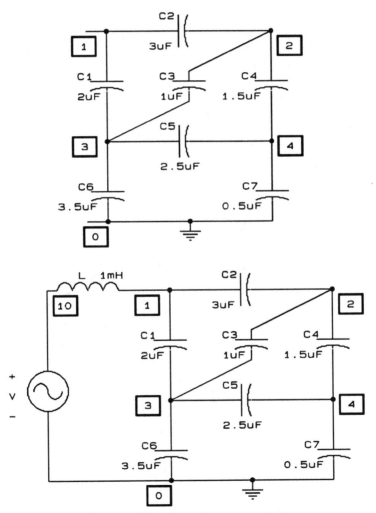

Figure 5.5 Circuits for Prob. 5.

5. An array of 7 capacitors is shown in Fig. 5.5.   The effec-
tive capacitance between nodes 1 and 0 is the unknown.
    a. Using the test circuit with an AC source and a 1 mH in-
    ductor shown, sweep the source over an appropriate range
    to make series resonance occur.   Look for resonance by ob-
    serving a plot of the voltage at node 1 (magnitude and
    phase) versus frequency.   HINT: PSpice requires that all
    nodes have a DC path to ground.   When nodes do not, you
    have to put in a dummy resistor (1E9 ohm will do) in order
    to run an analysis.
    b. As in Prob. 4.b, narrow the sweep range in the region
    of resonance to get an accurate measure of the resonant
    frequency.   Calculate effective capacitance using this
    result.

6. A 2-pole high-pass filter is shown in Fig. 5.6.   Its response is very under-damped, due to excessive gain.   In fact,
just a bit more gain and it will be a free-running oscillator.
    a. Model the op-amp as a 1E6 ohm input resistance and a
    VCVS with a gain of 5E4 V/V.   Make the input voltage be a
    pulse that goes from 0 V to 1 V at 0.5 ms, has a pulse
    width of 0.4 ms, and rise and fall times of 1E-6 s.   Write
    a PSpice input file that will do a transient analysis with

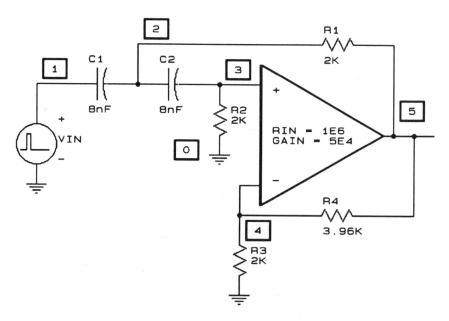

Figure 5.6 Circuit for Prob. 6.

Types of Analyses   Chap. 5

time steps of 0.1 ms up to 5 ms. Plot the output voltage
(node 5) and the input voltage (node 1) versus time. The
damped oscillation of the filter at its natural frequency
should be very apparent.

7. The Fourier analysis capability of Pspice can be used to
generate harmonic information for any periodic waveform. Figure 5.7 shows a 1 KHz square wave voltage source, V, with a 1
ohm resistor to ground to satisfy the requirement that each
node must be connected to at least two elements. The squarewave goes between -1 V and 1 V, and has rise and fall times of
1 ns.

    a. Write an input file using a PULSE function for V. Do a
transient analysis for 2 ms, with time steps of 1E-5 s.
Use the .FOUR control line to produce the Fourier components at node 1.

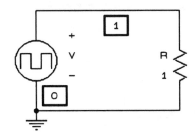

Figure 5.7 Circuit for Prob. 7.

Chapter 6

FORMATTING PSpice OUTPUT: PRINTING TABLES AND GRAPHS

6.1 INTRODUCTION

As we saw in Chap. 5 on the types of analyses possible with PSpice, some control lines will cause an automatic printing of analysis results. Among these control lines are: .OP (operating point), .TF (transfer function) and .SENS (sensitivity). Other kinds of analysis, such as .DC (non-linear DC), .AC (small-signal AC), .TRAN (non-linear transient analysis) and .NOISE (noise) do require that an additional control line be used to send results to the output file.

This chapter will cover two control lines, .PRINT and .PLOT, which, when additional parameters are included, will cause the analysis results to be written to the PSpice output file. .PRINT presents analysis results in tabular form, while .PLOT creates rectangular graphs of results. By investing a modest amount of time learning how these two control lines work, you will be able to make the output appear the way you want it to. Much time can be saved by letting PSpice do some of the major number crunching and then printing the final result. For example, PSpice can print and plot magnitudes in decibel form, as well as expressing complex numbers in both polar and rectangular forms.

6.2 USE OF THE .PRINT CONTROL LINE

6.2.1 .PRINT Format

The control line format is

.PRINT ANALTYPE OV1 <OV2 ... OV8>

ANALTYPE is one of four types of analysis:

1. DC
2. AC
3. TRAN
4. NOISE

Notice that the ANALTYPE does not have a period (.) preceding it. Up to 8 output variables can be listed together on one print control line, although only 5 output variables will fit on an 80 column-wide printer. There is no limit to the number of print control lines in an input file.

OV stands for output variable. The format for OV1 through OV8 depends on the type of analysis being done. The .PRINT line will make a table of results appear in the output file. The first column will be the DC voltage (for DC analysis), the frequency (for AC analysis) or the time (for transient analysis).

6.2.2 DC and Transient Analysis Output Variable Formats

Output variables for DC and transient analysis can be voltages at nodes compared to ground (node 0), voltages between two nodes or currents flowing through independent voltage sources. If an output variable has only one node listed, such as V(8), PSpice interprets that to be compared to ground.

EXAMPLES:

.PRINT DC V(1) V(5,7) I(VGATE)

The results of the DC analysis will be the magnitude of node 1 voltage (compared to ground), the voltage between nodes 5 and 7, and the current flowing through voltage source VGATE. The output data table will have four columns, one for the first DC source listed in the .DC control line and three for the three output variables.

.PRINT TRAN V(6) I(VSENSING) V(4,2)

The results of the transient analysis will be the magnitude of node 6 voltage, the voltage between nodes 4 and 2, and the current flowing through voltage source VSENSING. The output data table will have four columns, one for the time steps specified in the .TRAN control line and three for the three output variables.

Remember that PSpice uses dead voltage sources as ammeters, so an appropriate name for such a current-sensing dead voltage source is VSENSING.

6.2.3 AC Analysis Output Variable Formats

Output variables for AC analysis can be voltages at nodes compared to ground (node 0), voltages between two nodes, or currents flowing through independent voltage sources.

For AC analyses, the output variables in the .PRINT control line could be the same as in both examples above for DC and transient analysis. If this were done, PSpice would print only the magnitude of each voltage or current. Since sinusoidal voltages and currents can be complex, PSpice allows for printing complex voltages as follows:

V	magnitude
VR	real part
VI	imaginary part
VM	magnitude; V gives the same result.
VP	phase
VDB	20*LOG(magnitude)
VG	group delay (-$d$phase/$d$frequency)

Similarly, complex current can be specified as follows

I(VXXXXXX)	magnitude
IR(VXXXXXX)	real part
II(VXXXXXX)	imaginary part
IM(VXXXXXX)	magnitude
IP(VXXXXXX)	phase
IDB(VXXXXXX)	20*LOG(magnitude)
IG(VXXXXXX)	group delay (-$d$phase/$d$frequency)

EXAMPLES:

.PRINT AC VM(3) VP(3) VR(3) VI(3) VDB(3)

This print control line will make PSpice create a table of the results of an AC analysis. The leftmost column will be all the frequencies specified by the .AC control line. The five columns to the right of the frequency column will contain the magnitude and phase of the voltage at node 3 (compared to node 0, or ground), the real and imaginary parts of the voltage at node 3, and the magnitude of the node 3 voltage in decibels. Thus, PSpice has presented the phasor voltage at node 3 in polar form, rectangular form, and in dB.

.PRINT AC IM(VSENSING) IP(VSENSING) IDB(VSHORT)

The magnitude and phase of the phasor current through voltage source VSENSING will be printed along with the magnitude of the current through voltage source VSHORT, expressed in decibels.

6.2.4 Noise Analysis Output Variable Formats

Output variables for noise analysis are structured differently from DC, AC and transient analysis. The kinds of output variables are

ONOISE   output noise
INOISE   equivalent input noise

and, associated with an output variable, any of the following may appear in parentheses:

R   real part
I   imaginary part
M   magnitude
P   phase
DB   20*LOG(magnitude)

If an output variable appears without R, I, M, P, or DB in parentheses then the magnitude of that output variable will be printed.

EXAMPLE:

.PRINT NOISE INOISE ONOISE(DB)

Based on the noise analysis, the magnitude of the equivalent input noise and the output noise, in dB, will be printed at each frequency specified by the .NOISE and .AC control lines.

6.3 USE OF THE .PLOT CONTROL LINE

In many instances during the circuit design process, exact values of circuit voltages and currents are not needed. Very useful information may be obtained by looking at a graph of those parameters versus voltage (DC analysis), time (transient analysis) or frequency (AC analysis). Attempting to determine qualitative information, such as ripple in a Bode plot of filter gain, by examining tabular data can be time-consuming and ineffective. The .PLOT control line instructs PSpice to graph output data.

The plotting control line (.PLOT) follows the identical format as the print control line (.PRINT) except that the low and high plot limits can be specified for one or a group of output variables. If the optional plot limits are not specified, PSpice will automatically determine the maximum and minimum values of each variable being plotted, and will scale the output variable axis accordingly. The axis may be made linear or logarithmic automatically by PSpice. Although the automatic scaling results are not always to one's liking, it is a good idea to not specify plot limits the first time an analysis is run. Then one can examine the plot in the output file, choose appropriate plot limits and re-run the analysis with a revised input file.

PSpice uses a different symbol for each curve plotted. When two or more curves intersect, the plot symbol is a letter X at that point. The first output variable listed in a .PLOT control line will have its values printed in addition to being plotted. If you want the value of more than one output variable to be printed, a separate .PRINT control line can be used.

6.3.1 .PLOT Format

The control line format is

.PLOT ANALTYPE OV1 <(PLO1,PHI1)> <OV2 <(PLO2,PHI2)>
+ ... OV8 <(PLO8,PHI8)>>

Notice that the ANALTYPE again does not have a period (.) preceding it. Up to 8 output variables can be listed together on one plot control line. To avoid very confusing graphs that are difficult to interpret, it is best to restrict the number of variables on one .PLOT line to two or three. There is no limit to the number of .PLOT control lines in an input file.

OV stands for output variable; the format for OV1 through OV8 depends on the type of analysis being done. The .PLOT line will make a graph of results appear in the output file. The abscissa, or horizontal axis, will be the DC voltage (for DC analysis), the frequency (for AC analysis) or the time (for transient analysis).

(PLO, PHI) are plot limits which are optional and can be put (with parentheses) after any of the output variables. If plot limits are used, they cause all output variables to their left (up to the previous output variable whose plot limits are specified) to be plotted using PLO as the lowest axis value and PHI as the highest axis value.

If the plotted values have sufficiently different ranges, PSpice will use and label more than one ordinate scale unless optional plot limits override this.

6.3.2 DC and Transient Analysis Output Variable Formats

Output variables for DC and transient analysis can be voltages at nodes compared to ground (node 0), voltages between two nodes, or currents that flow through independent voltage sources.

EXAMPLE:

.PLOT DC V(1) (-4,12) V(5,7) I(VGATE)

The plotted results of the DC analysis will be the magnitude of node 1 voltage, the voltage between nodes 5 and 7, and the current flowing through voltage source VGATE. The output graph will have three plots; the abscissa (horizontal axis) for all three will be the first DC source listed in the .DC control line. The plots will be the voltage at node 1, with scale limits of -4 V to +12 V, the voltage between nodes 5 and 7, and the current through voltage source VGATE. The last two plots will be scaled automatically by PSpice.

EXAMPLE:

.PLOT TRAN V(6) I(VSENSING) V(4,2) (0,8)

The plotted results of the transient analysis will be the magnitude of node 6 voltage, the current flowing through voltage source VSENSING, and the voltage between nodes 4 and 2. The horizontal axis for the plots of the three output variables will be the time as specified in the .TRAN control line. All three plots will have vertical scale limits of 0 and 8; V(6) and V(4,2) will be in volts, ISENSING will be in amps.

6.3.3 AC Analysis Output Variable Formats

Output variables for AC analysis can be voltages at nodes compared to ground (node 0), voltages between two nodes, or currents flowing through independent voltage sources.

For AC analyses, the output variables in the .PLOT control line could be the same as in both examples above for DC and transient analysis. If this were done, PSpice would plot only the magnitude of each voltage or current. Since sinusoidal voltages and currents can be complex, PSpice allows for plotting complex voltages as follows:

V	magnitude
VR	real part
VI	imaginary part
VM	magnitude; V gives the same result.
VP	phase
VDB	20*LOG(magnitude)
VG	group delay (-$d$phase/$d$frequency)

Similarly, complex current can be specified as follows:

I(VXXXXXX)	magnitude
IR(VXXXXXX)	real part
II(VXXXXXX)	imaginary part
IM(VXXXXXX)	magnitude
IP(VXXXXXX)	phase
IDB(VXXXXXX)	20*LOG(magnitude)
IG(VXXXXXX)	group delay (-$d$phase/$d$frequency)

If the AC input voltage to a circuit has a magnitude of 1 V, use of VDB for an output voltage will produce a Bode plot, providing that a logarithmic sweep of frequency is done.

EXAMPLE:

.PLOT AC VM(3) VP(3)

This plot control line will make PSpice create a graph of the results of an AC analysis. The abscissa will be frequency, as specified by the .AC control line. The two plots will be the magnitude and phase of the voltage at node 3 (compared to node 0, or ground). The two vertical axes of the graph will be scaled automatically.

EXAMPLE:

.PLOT AC IM(VSENSING) (0,10M) VDB(35,19)

The magnitude of the phasor current through voltage source VSENSING will be plotted along with the magnitude of the voltage between nodes 35 and 19, expressed in decibels. The scale of the current plot will be 0 to 1E-02, while the voltage plot will be scaled automatically.

6.3.4 Noise Analysis Output Variable Formats

Output variables for noise analysis are structured differently from DC, AC and transient analysis. The kinds of output variables are:

ONOISE   output noise
INOISE   equivalent input noise

Associated with an output variable any of the following may appear, after the variable, in parentheses:

R   real part
I   imaginary part
M   magnitude
P   phase
DB   20*LOG(magnitude)

If an output variable appears without R, I, M, P or DB in parentheses then the magnitude of that output variable will be plotted.

EXAMPLE:

.PLOT NOISE INOISE ONOISE(DB)

Based on the noise analysis, the magnitude of the equivalent input noise and the output noise, in dB, will be plotted versus frequency as specified by the .NOISE and .AC control lines.

6.4 INTERPRETING PSpice OUTPUT INFORMATION

6.4.1 Interpreting .PRINT Numerical Data

The .PRINT control line creates tables of numerical information. Results are printed in scientific notation. For example, the number zero would appear in a table as .000E+00, where E indicates exponent of 10. So 7.345E+01 means 73.45, and 1.286E-02 means 0.01286. Be careful when reading results to avoid errors that can be off by one or more orders of magnitude. It is not hard to mistakenly interpret a current of 5.400E-7 A as 54 uA, when in fact it is 0.54 uA.

The example in Section 6.4.3 will illustrate how to use the .PRINT control line to create a table of results.

6.4.2 Interpreting .PLOT Numerical Data

The .PLOT control line creates a graph of numerical information. The first variable listed in a .PLOT line also has its value printed in scientific notation, with E meaning exponent of 10. It would be redundant to have the following two control lines

```
.PRINT    AC  VM(8,3)
.PLOT     AC  VM(8,3)
```

The .PRINT line will create a table with two columns: frequency (as specified in the .AC control line) and magnitude of the voltage between nodes 8 and 3. The .PLOT line will create those same two columns as well as plotting VM(8,3) versus frequency. Thus, the same information could be obtained with the .PLOT line alone.

When more than one variable is plotted, the .PLOT line will create only two columns. In this situation a .PRINT line could be used to print values of all data needed.

6.4.3 Example of .PRINT and .PLOT

The circuit in Fig. 6.1 shows a triangle voltage source connected to a resistor in series with an inductor. The triangle

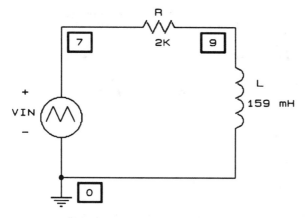

Figure 6.1 Series R-C Circuit with Triangle Input Voltage.

waveform is made by using an independent voltage source PULSE function, with the pulse width essentially zero. Figure 6.2 shows the input file LR.CIR, with the rise time set to 0.5 ms, fall time of 0.5 ms, and pulse width of 1 ns (since PSpice may give erroneous results if zero is used). Section 12.4 shows how to make a triangle waveform with the piecewise linear (PWL) source.

The control lines in LR.CIR instruct PSpice to print and plot the results of the transient analysis over two periods of the input voltage (a period is 1 ms). The .PRINT line causes a table to be made of all possible voltages. The first .PLOT control line has plot limits of 0 and 1 V for V(7). The second .PLOT line does not specify plot limits; PSpice will determine the scale automatically for V(9). The third .PLOT line will graph both parameters, and PSpice will determine the scale automatically for each plot.

Figure 6.3 is the output file LR.OUT, which contains a table and three graphs. The table has four columns: time, and

```
LR.CIR  TRIANGLE WAVEFORM MADE WITH PULSE FUNCTION, L-R HP FILTER
VIN     7  0  PULSE(0  1  0  .5M  .5M  1N 1M)
*       THE LINE ABOVE IS A 1 VPP TRIANGLE AT 1KHZ
R       7  9  2K
L       9  0  159M
.TRAN  50U  1.5M
.PRINT  TRAN  V(7)  V(9)  V(7,9)
.PLOT TRAN V(7)  (0,1)
.PLOT TRAN V(9)
.PLOT TRAN V(9) V(7)
.OPTIONS NOPAGE
.END
```

Figure 6.2 Input File LR.CIR.

```
**** 08/03/91 20:45:07 ********* Evaluation PSpice (January 1991) ************
LR.CIR  TRIANGLE WAVEFORM MADE WITH PULSE FUNCTION, L-R HP FILTER

****      CIRCUIT DESCRIPTION
*******************************************************************************

VIN     7  0  PULSE(0  1  0  .5M  .5M  1N  1M)
*       THE LINE ABOVE IS A 1 VPP TRIANGLE AT 1KHZ
R       7  9  2K
L       9  0  159M
.TRAN  50U  1.5M
.PRINT  TRAN  V(7)  V(9)  V(7,9)
.PLOT TRAN V(7)  (0,1)
.PLOT TRAN V(9)
.PLOT TRAN V(9) V(7)
.OPTIONS NOPAGE
.END

****      INITIAL TRANSIENT SOLUTION        TEMPERATURE =   27.000 DEG C

 NODE   VOLTAGE      NODE   VOLTAGE      NODE   VOLTAGE      NODE   VOLTAGE
(   7)   0.0000  (   9)    0.0000

       VOLTAGE SOURCE CURRENTS
       NAME           CURRENT
       VIN            0.000E+00

       TOTAL POWER DISSIPATION    0.00E+00  WATTS

****      TRANSIENT ANALYSIS              TEMPERATURE =   27.000 DEG C

   TIME         V(7)         V(9)         V(7,9)
  0.000E+00    0.000E+00    0.000E+00    0.000E+00
  5.000E-05    1.000E-01    7.417E-02    2.583E-02
  1.000E-04    2.000E-01    1.136E-01    8.638E-02
  1.500E-04    3.000E-01    1.352E-01    1.648E-01
  2.000E-04    4.000E-01    1.465E-01    2.535E-01
  2.500E-04    5.000E-01    1.523E-01    3.477E-01
  3.000E-04    6.000E-01    1.555E-01    4.445E-01
  3.500E-04    7.000E-01    1.572E-01    5.428E-01
  4.000E-04    8.000E-01    1.580E-01    6.420E-01
  4.500E-04    9.000E-01    1.585E-01    7.415E-01
  5.000E-04    1.000E+00    1.587E-01    8.413E-01
  5.500E-04    9.000E-01    1.209E-02    8.879E-01
  6.000E-04    8.000E-01   -6.948E-02    8.695E-01
  6.500E-04    7.000E-01   -1.112E-01    8.112E-01
  7.000E-04    6.000E-01   -1.339E-01    7.339E-01
  7.500E-04    5.000E-01   -1.459E-01    6.459E-01
  8.000E-04    4.000E-01   -1.520E-01    5.520E-01
  8.500E-04    3.000E-01   -1.553E-01    4.553E-01
  9.000E-04    2.000E-01   -1.571E-01    3.571E-01
  9.500E-04    1.000E-01   -1.580E-01    2.580E-01
  1.000E-03    3.958E-06   -1.585E-01    1.585E-01
  1.050E-03    1.000E-01   -1.129E-02    1.113E-01
  1.100E-03    2.000E-01    6.899E-02    1.310E-01
  1.150E-03    3.000E-01    1.112E-01    1.888E-01
  1.200E-03    4.000E-01    1.340E-01    2.660E-01
  1.250E-03    5.000E-01    1.458E-01    3.542E-01
  1.300E-03    6.000E-01    1.520E-01    4.480E-01
  1.350E-03    7.000E-01    1.553E-01    5.447E-01
  1.400E-03    8.000E-01    1.571E-01    6.429E-01
  1.450E-03    9.000E-01    1.580E-01    7.420E-01
  1.500E-03    1.000E+00    1.585E-01    8.415E-01
```

Figure 6.3 Output File LR.OUT.

the three voltages specified. The first graph shows the triangle voltage, V(7), starting at 0 V at 0 ms, rising to 1 V at
500 us and falling back to 0 V at 1 ms. The triangle is repeated from 1 ms to 2 ms. Notice that the magnitude of V(7)
is printed next to each time value along the time axis of the
graph.

The second graph shows the voltage across the inductor,
V(9). PSpice has scaled the voltage axis with good results.
The third graph shows both V(9) and V(7), for which the plot
symbols are asterisks (*) and plus signs (+) respectively.
Notice that only the magnitude of V(9) is printed along the
time axis. Each variable has its own voltage axis scale. Also, the automatic voltage axis scaling of V(7) results in a
compressed plot which is not as useful as the plot of V(7) in

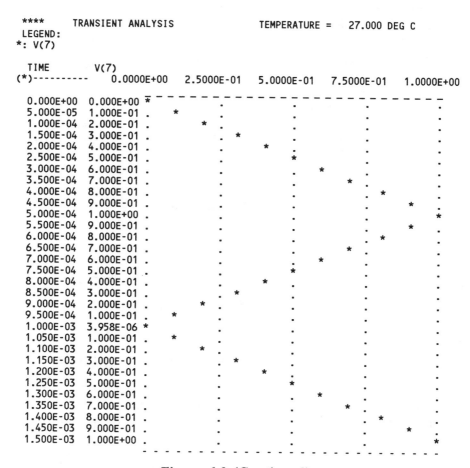

Figure 6.3 (Continued).

```
****    TRANSIENT ANALYSIS                   TEMPERATURE =    27.000 DEG C
 LEGEND:
 *: V(9)

   TIME         V(9)
 (*)----------    -2.0000E-01  -1.0000E-01   0.0000E+00   1.0000E-01   2.0000E-01

 0.000E+00   0.000E+00 .  - - - - - - - - - - - - - *  - - - - - - - - - - - -
 5.000E-05   7.417E-02 .                 .            .            *   .            .
 1.000E-04   1.136E-01 .                 .            .            .  *          .
 1.500E-04   1.352E-01 .                 .            .            .     *        .
 2.000E-04   1.465E-01 .                 .            .            .       *      .
 2.500E-04   1.523E-01 .                 .            .            .        *     .
 3.000E-04   1.555E-01 .                 .            .            .        *     .
 3.500E-04   1.572E-01 .                 .            .            .        *     .
 4.000E-04   1.580E-01 .                 .            .            .         *    .
 4.500E-04   1.585E-01 .                 .            .            .         *    .
 5.000E-04   1.587E-01 .                 .            .            .          *   .
 5.500E-04   1.209E-02 .                 .            .    *       .            .
 6.000E-04  -6.948E-02 .                 .      *     .            .            .
 6.500E-04  -1.112E-01 .               *.           .            .            .
 7.000E-04  -1.339E-01 .            *    .            .            .            .
 7.500E-04  -1.459E-01 .          *      .            .            .            .
 8.000E-04  -1.520E-01 .        *        .            .            .            .
 8.500E-04  -1.553E-01 .        *        .            .            .            .
 9.000E-04  -1.571E-01 .        *        .            .            .            .
 9.500E-04  -1.580E-01 .      *          .            .            .            .
 1.000E-03  -1.585E-01 .      *          .            .            .            .
 1.050E-03  -1.129E-02 .                 .          *.            .            .
 1.100E-03   6.899E-02 .                 .            .       *    .            .
 1.150E-03   1.112E-01 .                 .            .          .*          .
 1.200E-03   1.340E-01 .                 .            .            .    *       .
 1.250E-03   1.458E-01 .                 .            .            .      *     .
 1.300E-03   1.520E-01 .                 .            .            .       *    .
 1.350E-03   1.553E-01 .                 .            .            .       *    .
 1.400E-03   1.571E-01 .                 .            .            .       *    .
 1.450E-03   1.580E-01 .                 .            .            .         *  .
 1.500E-03   1.585E-01 .  - - - - - - - - - - - - - - - - - - - - - -*- - - - -
```

Figure 6.3 (Continued).

the first graph.   This could be improved by changing the third
.PLOT line in Fig. 6.2 to

.PLOT TRAN V(9) (-0.2, 0.2) V(7) (0,1)

Recall that plot limits affect all output variables to the
left of the plot limit specification, unless different plot
limits were specified.   Thus, if the plot limit (-0.2, 0.2)
was not present, both V(9) and V(7) would use the (0,1) plot
limit.   This would chop off the bottom part of the V(9) waveform.

```
 ****   TRANSIENT ANALYSIS              TEMPERATURE =   27.000 DEG C
  LEGEND:
 *: V(9)
 +: V(7)

  TIME         V(9)
  (*)---------   -2.0000E-01  -1.0000E-01  0.0000E+00  1.0000E-01  2.0000E-01
  (+)---------   -5.0000E-01   0.0000E+00  5.0000E-01  1.0000E+00  1.5000E-01

0.000E+00  0.000E+00 . - - - - - - - + - - - - - * - - - - - - - - - - - - -
5.000E-05  7.417E-02 .             .    +        .         *        .
1.000E-04  1.136E-01 .             .        +    .              . *
1.500E-04  1.352E-01 .             .           + .                *
2.000E-04  1.465E-01 .             .              + .               *
2.500E-04  1.523E-01 .             .               +.                 *
3.000E-04  1.555E-01 .             .                . +                *
3.500E-04  1.572E-01 .             .                .   +              *
4.000E-04  1.580E-01 .             .                .     +            *
4.500E-04  1.585E-01 .             .                .       + .        *
5.000E-04  1.587E-01 .             .                .         +.       *
5.500E-04  1.209E-02 .             .         *      .        +  .
6.000E-04 -6.948E-02 .             .   *            .      +    .
6.500E-04 -1.112E-01 .          *.                  .    +
7.000E-04 -1.339E-01 .       *    .                 .   +
7.500E-04 -1.459E-01 .     *      .                 .  +
8.000E-04 -1.520E-01 .    *       .              +  .
8.500E-04 -1.553E-01 .    *       .            +    .
9.000E-04 -1.571E-01 .    *       .         +      .
9.500E-04 -1.580E-01 .   *        .     +    .
1.000E-03 -1.585E-01 .   *        +          .
1.050E-03 -1.129E-02 .            .     +    *.
1.100E-03  6.899E-02 .            .        +        *
1.150E-03  1.112E-01 .            .           +    .          .*
1.200E-03  1.340E-01 .            .              + .              *
1.250E-03  1.458E-01 .            .               +.               *
1.300E-03  1.520E-01 .            .                . +              *
1.350E-03  1.553E-01 .            .                .   +            *
1.400E-03  1.571E-01 .            .                .     +          *
1.450E-03  1.580E-01 .            .                .        +       *
1.500E-03  1.585E-01 .            .                .          +     *
                     . - - - - - - - - - - - - - - . - - - - - - - - - - - - -

          JOB CONCLUDED
          TOTAL JOB TIME            5.66
```

Figure 6.3 (Continued).

6.4.4   Obtaining Higher-Quality Graphs

As we can see from Fig. 6.3, graphs done on a line printer using punctuation symbols leave something to be desired. While PSpice itself cannot directly produce high-quality graphs, MicroSim Corporation offers a graphics post-processor program called Probe. PSpice can be told to write the results of an analysis to a data file which Probe will use to produce high quality graphical output. Chapter 7 is devoted to Probe.

In addition to the professional-looking graphs that result, Probe can perform mathematical operations on output variables and graph the results.

CHAPTER SUMMARY

Results of DC, AC, transient and noise analyses can be printed in tabular format and/or plotted by PSpice.

Complex AC analysis results can be expressed in two ways: polar (VM and VP) and rectangular (VR and VI). Also the decibel expression of a voltage or current magnitude can be computed by PSpice.

Graphs will be scaled automatically by PSpice, or you can specify optional plot limits to make the graphs appear the way you want them.

While the graphs produced by PSpice are very useful, a graphics post-processor (Probe) which produces professional-looking results is available from MicroSim Corporation.

Chapter 7

Probe - A GRAPHICS POST PROCESSOR

7.1 INTRODUCTION

SPICE is capable of making graphs of analysis results. These graphs, however, are primitive, difficult to interpret, and can cause a huge amount of paper to be wasted when they are printed. Standard ASCII symbols (*, +, =) are used to make the graphs on a line printer. Further, these graphs can show only voltages and currents that result from the analysis done.

In contrast, a graphics post-processor called Probe uses the data produced by a PSpice analysis and can produce very clear, professional-looking graphs of analysis results. In addition, Probe can show graphs of mathematical functions of analysis results: a voltage divided by a current gives an impedance; a voltage multiplied by a current produces instantaneous power; a Fourier transform of a voltage plotted versus time gives a graph of voltage plotted as a function of frequency.

Probe runs after PSpice is done. It reads the PSpice analysis results from a data file (PROBE.DAT) and can display multiple plots (a plot is a separate graph with its own y-axis, but having the same x-axis as the other plots). On each plot there can be multiple traces (a trace is a graph of a variable, sharing both the x- and y-axis with other traces).

Once Probe has displayed the graphical data in the desired way, it can send the graphs to a wide range of printers and plotters.

7.2 DO THIS BEFORE USING PROBE

There are two things you must do to get your computer ready for running Probe. They are rather straightforward and simple to do; however, Probe will not run correctly until they are done.

1. The file PROBE.EXE must be present in the PSpice subdirectory in which you are working. To see if it is there, type DIR *.EXE while in your PSpice subdirectory. If PROBE.EXE is not there, copy it from a floppy disk to your hard disk PSpice subdirectory.

2. A file PROBE.DEV or PSPICE.DEV must be present in the PSpice subdirectory, and it <u>must</u> have been customized (by you) for your computer and its output devices. It tells Probe what type of graphics display your computer uses, what printer (or graphics plotter) is connected to the computer, and through which port (LPT1, COM2, etc.) it is connected. A typical PROBE.DEV file would look like:

> Display = IBMEGA
> Hard-copy = LPT1:, EPSON
> Hard-copy = COM2:, HP6

This file tells Probe that the display is an IBM Enhanced Graphics Adapter and that there are two printers connected to the computer: an Epson FX80 or similar dot-matrix printer connected to the first parallel printer port (LPT1:) and a Hewlett-Packard 6-pen graphics plotter connected to the second serial port (COM2:).

A text editor (that works with, and produces, ASCII files) can be used to alter the file PROBE.DEV to make it correctly describe your computer and its printer(s) to Probe. Allowed display types (and resolutions) for the IBM-PC include:

AT&T   Standard AT&T 6300, monochrome,
                             640x400

GenericEGA	Enhanced Graphics Adapter, slow but compatible
Hercules	Hercules Graphics Card, monochrome, 720x348
IBM	Color Graphics Adapter, monochrome, 640x200
IBMClr	Color Graphics Adapter, 320x200
IBMEGA	Enhanced Graphics Adapter, 640x350
IBMEGAMono	Enhanced Graphics Adapter, monochrome, 640x350
IBMMCGA	for IBM PS/2 model 30, monochrome, 640x480
IBMVGA	Video Graphics Adapter, 640x480
Text	for non-graphic displays - can be used with any display
T3100	Toshiba 3100 laptop, monochrome, 640x400

For a complete listing of displays supported by Probe, see the PSpice User's Guide for your version of PSpice and Probe. If you find that your display type is not listed, you can use trial and error until you find one that works for your system. If you specify the wrong kind, you may find that your display goes blank when you enter Probe; if so, exit Probe and edit PROBE.DEV to try another display type. While the "Text" display type will work for all displays, you can't realize the full benefits of Probe with it.

Allowed port names for the IBM-PC are COM1, COM2, FILE, LPT1, LPT2, LPT3, and PRN. There are many printers that are supported by PSpice; among the more common types are:

EPSON	Epson FX80 or FX100 dot-matrix
EPSONLQ	Epson LQ series, 24-pin dot-matrix
HI	Houston Instrument DMP pen plotter
HP	Hewlett-Packard 7400 and 7500 pen plotter family
HP6	Hewlett-Packard six-pen pen plotters
HP8	Hewlett-Packard 7550A eight-pen pen plotter
HPLJ	Hewlett-Packard 2686A, PLUS, or Series II LaserJet, 75 dots/inch, serial or parallel interface can be used
HPLJ100	Hewlett-Packard PLUS or Series II LaserJet, 100 dots/inch, serial or parallel (preferred) interface
HPLJ150	Hewlett-Packard PLUS or Series II LaserJet, 150 dots/inch, serial or parallel (preferred) interface

| HPLJ300 | Hewlett-Packard PLUS or Series II LaserJet, 300 dots/inch, serial or parallel (preferred) interface |
| TEXT | Non-graphic printer, uses ASCII characters to make plots |

For a complete listing of printers supported by Probe, see the PSpice User's Guide for your version of PSpice and Probe.

7.3 FIRST EXAMPLE - TRANSIENT ANALYSIS

Figure 7.1 is a series resonant circuit, fed by a sinusoidal voltage source at the circuit's resonant frequency of 10 KHz. The PSpice input file, in Fig. 7.2, describes the circuit and contains the control line ".PROBE". This instructs PSpice to write the results of the transient analysis to a file called PROBE.DAT. When the analysis has been completed, the batch file PSPICE.BAT will look for the file PROBE.DAT in the default directory. If PROBE.DAT exists, then the batch file will run PROBE.EXE, the executable file which reads PROBE.DAT and can present the data in PROBE.DAT in a wide variety of formats. Notice the complete absence of .PRINT and .PLOT control lines in the input file. A .PRINT line would give numerical data which are difficult to interpret, while a .PLOT line would produce a crude graph in the PSpice output file which is made obsolete by the graphics post-processing capabilities of Probe.

Figure 7.3 is a graph of three voltages in the circuit on a single plot: the source voltage V(1) (a 10 KHz, 5 Vp sinusoid), and the voltages across the inductor, V(1,2), and the capacitor, V(2,3). Since there are different magnitudes and phases involved, the graph (plot) is a little crowded and dif-

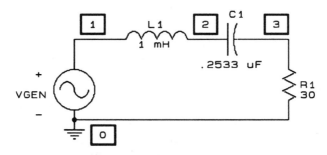

Figure 7.1 Series Resonant Circuit.

```
SER-RES.CIR    SERIES RESONANT CIRCUIT, 10 KHZ
VGEN  1  0  SIN(0  5  10KHz)
L1  1  2  1M
C1  2  3  0.2533U
R1  3  0  30
.TRAN  5U  300U  0  2U
.PROBE
.END
```

Figure 7.2 Input File SER-RES.CIR.

ficult to interpret.   Figure 7.4 contains two plots: the upper one is just the source voltage by itself, while the lower plot, with two traces, shows both the inductor and capacitor voltages.   The upper one has different axis scaling compared to the lower plot; this allows all voltages to be shown more clearly, including the phase relationships between voltages.

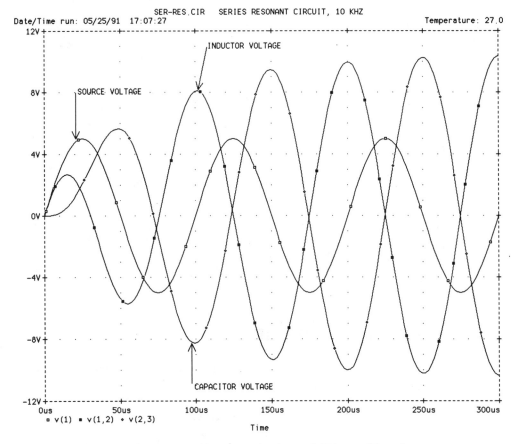

Figure 7.3 Probe Display of Three Circuit Voltages.

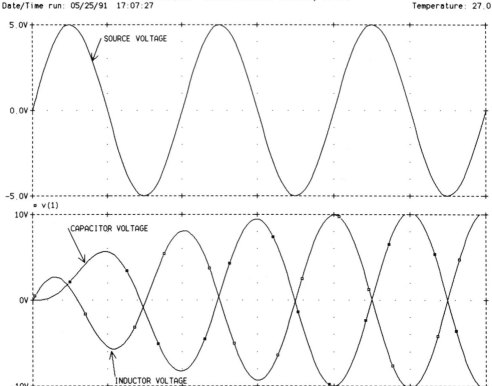

Figure 7.4 Probe Dislay Using Two Plots.

The voltage across the inductor, V(1,2), leads the input voltage by 90 degrees, and the capacitor voltage, V(2,3) is seen to lag the input by 90 degrees. The source voltage is 5 Vp, while the inductor and capacitor voltages build up with time to about 11 Vp.

7.4 MENUS WITHIN PROBE

When Probe is run, the first menu you see depends on whether more than one type of analysis was done. If there were two or more, Probe will begin with a start-up menu which asks you to select which (DC, AC, transient) analysis to display. If only one analysis was done, the screen has a blank graph at the top, with a menu (called the Analog Plot Menu) at the bottom containing the following choices:

Exit, Add_trace, X_axis, Y_axis, Plot_control,
Display_control, Macros, Hard_copy, Zoom, Label.

To select one of these menu items, press the capitalized letter in that item. For example, to make a graph of the voltage
at node 2, first press the letter "A", then key in V(2,0) when
prompted and press the ENTER key. This will cause a "trace"
of the voltage between node 2 and node 0 to be "added" to the
blank graph.
    After one or more traces have been added, a slightly dif-
ferent Analog Plot Menu appears:

Exit, Add_trace, Remove_trace, X_axis, Y_axis,
Plot_control, Display_control, Macros, Hard_copy, Cursor,
Zoom, Label.

This new menu has two additional items: Remove_trace and Cursor, which are useful only if there is a trace on the plot.
See Fig. 7.5, the screen seen within Probe while Fig. 7.4 was
being made. Labels were added to help make the plots clearer.

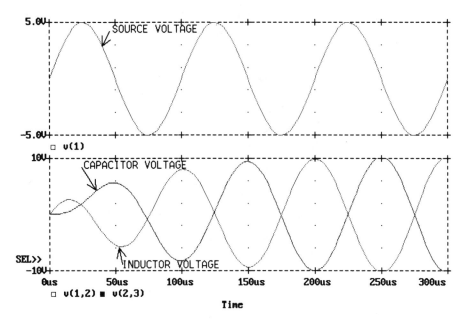

Figure 7.5 Analog Plot Menu in Probe.

The purpose of each item in this menu is as follows:

Exit - (or pressing the ESCape key) returns you to the Start-up Menu. Be careful, because if you return to the start-up menu and then go back into Probe, you will have to repeat all the commands you gave in Probe to customize your display. See Display_control item below on how to save customized displays.

Add_trace - causes one or more traces to be put on the plot. Each trace can be an output variable or an arithmetic expression containing an output variable. A list of available variables can be obtained by pressing the F4 key. Note: the menu item Remove_trace will appear once an Add_trace has been done. Remove_trace allows deletion of one or more traces from a plot.

Mathematical operators are + (addition), - (subtraction), * (multiplication) and / (division). Parentheses may be used. The mathematical functions that can be used in arithmetic expressions include:

Function	Meaning of Function
ABS(Z)	absolute value of Z ($|Z|$)
SGN(Z)	+1 if Z > 0, 0 if Z = 0, -1 if Z < 0
SQRT(Z)	squareroot of Z
EXP(Z)	2.718282 raised to the Z power
LOG(Z)	natural (base e) log of Z
LOG10(Z)	(base 10) log of Z
DB(Z)	20*LOG10(Z)
PWR(Z,Y)	$|Z|$ raised to the Y power
SIN(Z)	sine of Z, Z is in radians
COS(Z)	cosine of Z, Z is in radians
TAN(Z)	tangent of Z, Z is in radians
ATAN(Z)	inverse tangent, result is in radians
ARCTAN(Z)	inverse tangent, result is in radians
d(Z)	derivative of Z with respect to the X-axis variable
s(Z)	integral of Z over the range of the X-axis variable
AVG(Z)	running average of Z over the range of the X-axis variable
RMS(Z)	running RMS average of Z over the range of the X-axis variable
MIN(Z)	minimum of Z over the range of the X-axis variable

MAX(Z)   maximum of Z over the range of the X-axis
variable

Remove_trace - allows one or all traces to be removed from the selected plot.

X_axis - produces the Axis menu, which allows changes to be made to the X axis of all plots simultaneously. Under the Axis menu, the X axis can be made logarithmic (or linear if it is logarithmic), can have its range set (minimum and maximum values) or can be autoranged, can have its variable changed (the default variable is time when a transient analysis is done, frequency when an AC analysis is done, and voltage or current for a DC analysis.

In addition, a Fourier analysis can be done (if the X axis is time, then the X axis is converted, via an FFT, to frequency; if the X axis is frequency, an inverse Fourier transform converts the X axis to time), and the Restrict data menu item can be used to set the range of X axis data for an FFT or inverse FFT. Once Fourier has been selected, the Fourier menu item is replaced by Quit_fourier, which returns the X axis to what is was before the Fourier menu item was selected. See Section 7.7 for an example using Fourier analysis.

Y_axis - produces the Axis menu, which allows changes to be made to the Y axis of a plot. Under the Axis menu, the Y axis can be made logarithmic (or linear if it is logarithmic), can have its range set (minimum and maximum values) or can be autoranged.

Plot_Control - produces the Plot Control Menu, which allows you to add (or remove) a plot (at the top of the screen), thus making existing plot(s) smaller in height. All plots will share the same X axis. In addition, one of the plots can be "selected", allowing other Analog Plot Menu commands to be used (such as Add_trace) on just the plot that is selected.

Display_control - produces the Display Control Menu, which makes it possible to save and restore attributes of the plots and traces displayed *without reference to any specific data file*. The plot and trace attributes are saved in a file called PROBE.DSP. For example, it is possible to save the following attributes: the upper plot has trace V(1), with particular range values for its Y axis, the lower plot has traces V(1,2) and V(2,3), with certain range values for its Y axis (and linear or logarithmic spacing). This allows the display file which has been saved to be used with any data file (which has the voltages V(1), V(1,2) and V(2,3)). In this way, the PSpice input file for a specific circuit can be changed repeatedly, and Probe, using the saved display file, will show the graphical results in the same way without having to go through all the steps each time.

NOTE: It is a good idea to rename the display file (Probe saves it as PROBE.DSP) to another name (e.g. SER-RES.DSP), so that it can be associated with a particular input circuit file (e.g. SER-RES.CIR). If you don't rename it, the next time you create a display file it will overwrite the first display file and you will lose it.

The steps to be followed to create a display file include:

a) Do a PSpice analysis, with input file FILENAME.CIR. In that file, have the control line .PROBE.
b) When the analysis is over, and Probe is run automatically, make the plot(s) exactly the way you want them (using choices in the Analog Plot Menu such as Add_trace, Y_axis, X_axis, etc.).
c) From the Analog Plot Menu choose Display_control.
d) The Display Control Menu will appear; choose Save, and give the display a name (e.g. NEATO) when prompted to do so.
e) At this point, you have created a file called PROBE.DSP which contains the attributes of the display you created. Exit Probe, and rename PROBE.DSP to FILENAME.DSP.

NOTE: It would also be wise at this time to rename PROBE.DAT, the file containing the results of the analysis done when FILENAME.CIR was run, to FILENAME.DAT.

Assuming that you now have a file called FILENAME.DAT containing the analysis results, and a separate file called FILENAME.DSP which has the desired display attributes, the steps to be followed to see the display again are:

a) Type in PROBE/S FILENAME.DSP FILENAME.DAT and press ENTER.
b) When the Analog Plot Menu appears, press D (thus selecting Display_control).

c) Highlight the name (e.g. NEATO) you gave the display you wish to see again and press ENTER.

The graphs and the way you displayed them before you saved the display are now shown again. Use of the Display_control menu item can save large amounts of time and is well worth mastering by the experienced user.

Macros - produces the Macro Control Menu, which allows you to define, modify, copy, list and delete macros. A macro is a statement which can define a mathematical relationship between variables like voltage and current, or can define a constant. Examples of macros are:

$$PWR(M,N) = M*N \qquad ADMITT(M,N) = N/M \qquad COUL(N) = s(N)$$

In the above macro definitions, presumably M would be a voltage and N would be a current, so that the first macro (named PWR) defines power as the product of voltage and current, the second macro (named ADMITT) defines admittance as current divided by voltage, and the third macro (named COUL) defines the charge, in coulombs, as the integral of current with respect to time. Actually, the macro COUL will integrate anything that is put in as an argument; if the macro were written COUL(V*I), it would integrate instantaneous power (V*I) with respect to time, and the resulting display would be a graph of energy.

Macros are stored in a file called PROBE.MAC and can be viewed or edited with an ASCII text editor. The macros stored in PROBE.MAC can be used by Probe with any data file, not just the one that was used to create the macro.

Hard_copy - allows you to make a copy of the Probe plots on a printer or plotter (which were defined previously by you in the file PROBE.DEV). When this choice is selected, the available hardcopy devices are shown and you may select one. If a printer is selected, you are offered the choice of number of pages (one or two) or a length in inches. If a graphics pen plotter is selected, you may choose the size of the paper (A, B, C or D) or the paper size in inches. While the hardcopy is being made, the computer will not respond to keyboard commands.

Cursor - this command allows you to use two cursors to get numeric information from the traces selected. You might want to measure the risetime of a pulse, the time delay between two points, or the bandwidth of a tuned circuit. Traces are selected as follows:

> **First cursor:** press <CTRL> and right or left arrows
> **Second cursor:** press <SHIFT><CTRL> and right or left arrows

Cursors are moved on a trace as follows:

> **First cursor:** press the right or left arrows to move right or left; press <HOME> to move to the beginning of a trace; press <END> to move to the end of a trace.
> **Second cursor:** the same as for the first cursor, except the <SHIFT> key is pressed at the same time.

Each key press in moving a cursor moves one pixel (picture element) on the display; holding the key(s) down moves the cursor repeatedly by 10 pixels.

In the cursor menu are choices that allow you to make a hardcopy (see Hard_copy paragraph above for details) and search through a displayed trace for a peak, trough, slope, minimum or maximum. In addition, complex search commands can be given; refer to the PSpice User's Guide for your version of Probe for specifics.

Zoom - allows you to magnify a region of the display by zooming in on it. You can specify a region to view, zoom in or zoom out in the x or y direction, pan in the x or y direction (move left-right or up-down), or autorange (return to the original display).

Label - this menu choice permits you to annotate each plot. You can add text, lines, arrows, boxes, circles and ellipses. Another feature is the ability to change the title (by default the title line from the PSpice input file is put on each hardcopy).

7.5 SECOND EXAMPLE - AC ANALYSIS

Probe can not only display voltages and currents, it can also be used to make graphs of mathematical expressions of output variables. In Fig. 7.6 a tank circuit is shown which is paral-

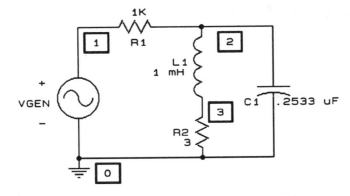

Figure 7.6 Parallel Resonant Circuit.

lel resonant at 10 KHz, fed by a sinusoidal voltage source in series with a resistor. The source frequency is varied in steps from 5 KHz to 15 KHz. The PSpice input file ACTANK.CIR is shown in Fig. 7.7.

For certain applications (such as tuned amplifiers) it would be useful to know the impedance of the tank circuit at frequencies below, at, and above resonance. Probe can be used to display a graph of impedance versus frequency by dividing the voltage across the tank, V(2,0), by the current through the tank, I(R1). Since both the current and voltage are phasor quantities, two plots will be needed: one for impedance magnitude and one for impedance angle. Figure 7.8 shows these two plots. In the upper one, the voltage magnitude, VM(2), is divided by the current magnitude, IM(R1), and in the lower plot the phase angle of the current is subtracted from the phase angle of the voltage to give the angle of the impedance.

```
ACTANK.CIR    PARALLEL RESONANT CIRCUIT, 10 KHZ
R1  1  2  1K
R2  3  0  3
VGEN 1 0 AC 1
L1  2  3  1M
C1  2  0  0.2533U
.AC LIN 21  5000 15000
.PRINT AC VM(2) VP(2)
.PLOT AC  VM(2) VP(2)
.END
```

Figure 7.7 Input File ACTANK.CIR.

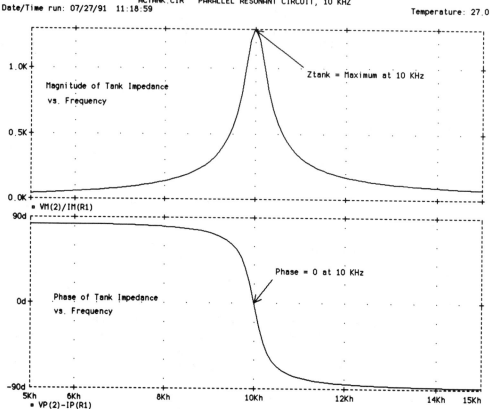

Ztank = Maximum at 10 KHz

Magnitude of Tank Impedance
vs. Frequency

▫ VM(2)/IM(R1)

Phase = 0 at 10 KHz

Phase of Tank Impedance
vs. Frequency

▫ VP(2)-IP(R1)

Frequency

**Figure 7.8 Probe Display of Tank
Impedance Magnitude and Phase.**

7.6 THIRD EXAMPLE - TRANSIENT ANALYSIS

It is well known that a DC voltage source connected to a series R-C circuit will produce an exponential growth of voltage
across the capacitor and an exponential decay of resistor voltage and current. Probe will be used here to illustrate an interesting characteristic of this circuit: the power supplied
by the source to the capacitor, which is the product of voltage across the capacitor and the capacitor current, reaches a
maximum during the first time constant and then decays to
zero.

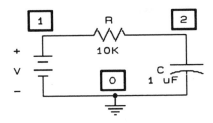

Figure 7.9 R-C Circuit.

Figure 7.9 shows the R-C circuit; what is assumed is that the capacitor is initially uncharged. RC-CHG.CIR, the PSpice input file, is shown in Fig. 7.10, and shows that the initial condition of the capacitor (IC = 0) is uncharged. The circuit time constant, R*C, is 10 ms, so the transient analysis is allowed to run for 50 ms, or five time constants.

The Probe two-plot display of the analysis results, Fig. 7.11, shows the voltage across the resistor and the capacitor in two traces of the upper plot. The lower plot shows the current through the capacitor on one trace, and the power delivered to the capacitor, V(2)*I(C), on the other trace. Using the cursor feature on the power trace, the peak power is found to be 2.500 mW, at 6.9 ms.

The time at which peak power is delivered to the capacitor in this kind of circuit will always be ln(2)*R*C, or 0.69*R*C. To prove this, multiply the expression for the capacitor voltage by the expression for the capacitor current, and take the derivative with respect to time. Set the derivative equal to zero, and solve for the time at which the maximum power delivered to the capacitor occurs.

```
RC-CHG.CIR    Capacitor Charging Through a Resistor
V  1  0  DC  10
R  1  2  10K
C  2  0  1E-6 IC=0
.TRAN  .05E-3  50E-3  0  .05E-3  UIC
.PROBE
.END
```

Figure 7.10 Input File RC-CHG.CIR.

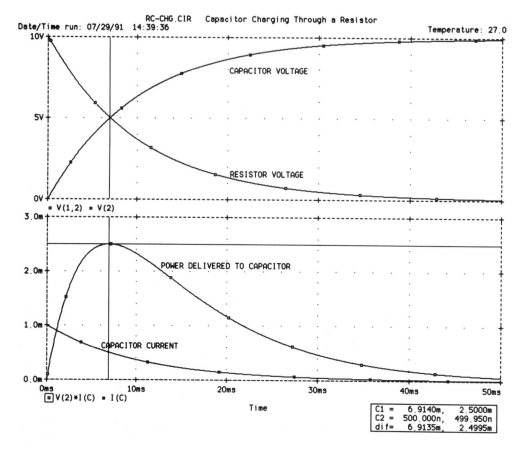

C1 =	6.9140m,	2.5000m
C2 =	500.000n,	499.950n
dif=	6.9135m,	2.4995m

Figure 7.11 Two-Plot Probe Display.

7.7 FOURTH EXAMPLE - FOURIER TRANSFORM

As mentioned in Section 7.4, the X_axis menu includes Fourier
as a menu item. A very simple circuit, Fig. 7.12, has a 1 KHz
squarewave generator connected to a 1-ohm resistor (the resis-

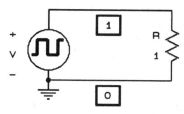

**Figure 7.12 Squarewave for Fourier
Analysis.**

```
SQUARE.CIR  1KHz squarewave, in time & frequency domain
V  1  0  PULSE(0  1  0  1N  1N  .5M  1M)
R  1  0  1
.TRAN  .05M  25M
.OPTIONS ITL5=0
.PROBE
.END
```

Figure 7.13 Input File SQUARE.CIR.

tor is needed to satisfy the requirement that at least two circuit elements must be connected to each node). The PSpice input file SQUARE.CIR, Fig. 7.13, describes the 1 KHz squarewave
as having risetime and falltime of 1 ns and a pulse width of
0.5 ms. The .OPTIONS ITL5=0 line allows the PSpice transient
analysis to run even though the default limit of 5000 itera-
tions is exceeded.

In Fig. 7.14, 25 periods of the squarewave are presented
in a graph of voltage vs. time. If the X axis menu choice

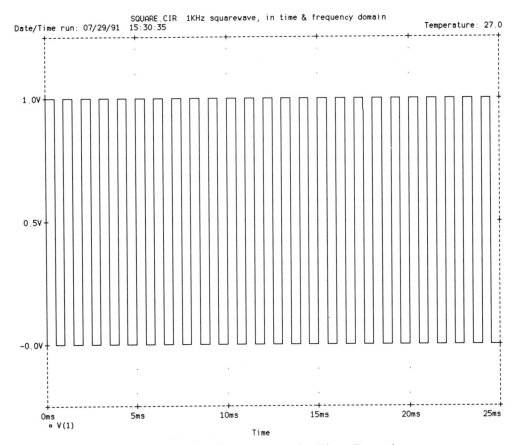

Figure 7.14 Squarewave in Time Domain.

Chap. 7 Probe - A Graphics Post Processor   **105**

Fourier is selected, a graph of voltage vs. frequency is created by Probe (Fig. 7.15), which shows that the 1 KHz squarewave
has only odd harmonics of 1 KHz. The 3 KHz, 5 KHz, 7 KHz and
9 KHz harmonics are displayed, with the Probe cursors indicating the amplitudes of the 1 KHz and 3 KHz harmonics. The amplitudes shown have an error of less than .05% compared to calculated harmonic amplitudes for a 1 Vpp squarewave.

One very important concept in using Probe for Fourier analysis is that the resolution of the voltage vs. frequency graph
is inversely proportional to the total time interval of the
voltage vs. time graph. Thus, it may be necessary to perform
the transient analysis for many periods of the input waveform
in order to get acceptably small resolution in the frequency
domain.

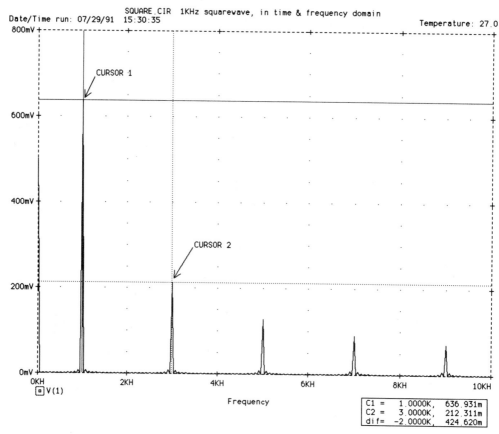

**Figure 7.15 Squarewave in Frequency
Domain.**

Probe - A Graphics Post Processor *Chap. 7*

CHAPTER SUMMARY

PSpice includes a graphics post-processor called Probe which allows the numerical results of circuit simulations to be displayed as high-quality graphs. These graphs can be printed or plotted. Probe has the capability to perform mathematical operations on voltages and currents, and can be used to graph power, impedance, or any quantities which can be expressed as functions of voltage and current.

Before using Probe, you must customize a file (PROBE.DEV) to indicate which display and hardcopy devices are used by your computer. Complex graphical results with multiple traces and plots can be saved, to be restored later. The Zoom feature allows precise viewing of any part of a graph, and the Cursor feature permits you to get precise numerical data from the graphs. Graphs in the time domain, via a fast Fourier transform, can be viewed in the frequency domain.

Chapter 8

SEMICONDUCTORS
IN PSpice

8.1 INTRODUCTION

PSpice has built-in models for four semiconductor devices: diodes, bipolar junction transistors (BJTs), field-effect transistors (FETs) and metal-oxide semiconductor field-effect transistors (MOSFETs). These internal models are typical of such
devices when created on an integrated circuit. Each type has
a large number of parameters, and PSpice has specific default
values for each parameter. For example, the PSpice model for
a BJT has 55 parameters including forward beta (DC current
gain), reverse beta (current gain when you interchange collector and emitter), emitter resistance, and a number of others
little known to most people who practice electronics. The default value (the value PSpice assigns to the parameter <u>unless
you specify otherwise</u>) for forward beta is 100, while reverse
beta defaults to 1.

  In this chapter the element lines for the four kinds of
semiconductor devices known by PSpice will be presented, along
with the .MODEL control line which must be used whenever a
semiconductor element line is included in the input file. Before you become overwhelmed with the prospect of modeling a
circuit with semiconductors, rest assured that most of the

time you need not deal with the vast majority of the myriad parameters. The only data needed to specify each semiconductor are a single element line and a single, short .MODEL line which specifies only those few parameters which you feel are important. In fact, any number of element lines for the same kind of semiconductor device (e.g., an NPN BJT transistor) can "share" the same single .MODEL line.

8.2 DESCRIBING DIODES TO PSpice

8.2.1 Diode Element Line

The general form for an element line for a junction diode is

DXXXXXX N+ N- MODNAME <AREA> <OFF> <IC=VD>

where DXXXXXX is the name of the diode, N+ is the node to which the anode is connected, N- is the node to which the cathode is connected and MODNAME is the model name which is used in an associated .MODEL control line. All of these are required in the diode element line. The optional parameters are:

AREA - the area factor which determines how many of the diode model MODNAME are put in parallel to make one DXXXXXX. The area parameter will affect IS, RS, CJO and IBV in the model of the diode (see Section 8.2.3). If not specified, area defaults to 1.

OFF - an initial condition of DXXXXXX for the DC analysis.

IC=VD - causes PSpice to use VD as the initial condition for diode voltage instead of the quiescent operating point diode voltage when a transient analysis is done.

8.2.2 Diode Model Line

The general form for a .MODEL line for a junction diode is

.MODEL MODNAME D<(PAR1=PVAL1 PAR2=PVAL2 ...)>

where MODNAME is the model name given to a diode in an element line, and D tells PSpice that the semiconductor being modeled is a diode and not a BJT, FET or MOSFET. PAR is the parameter name, which can be any of those in the parameter list in Section 8.2.3, and PVAL is the value of that parameter. Note that the PAR and PVAL parts of the .MODEL line are optional; the minimum specification of a diode is simply the letter D.

EXAMPLE:

DRECTIFY 7 15 HIPOWER
MODEL HIPOWER D(RS = 0.5 BV = 400 IBV = 50M)

The diode named DRECTIFY has its anode at node 7 and its cathode at node 15. Its model name is HIPOWER. The model named HIPOWER includes 0.5 ohm of ohmic resistance, a reverse breakdown voltage of 400 V, and a current at reverse breakdown of 50 mA.

EXAMPLE:

DBIG10 9 4 DHEFTY
DBIG20 5 9 DHEFTY 2.5
.MODEL DHEFTY D(IS = 2E-14 BV = 8E2 IBV = 0.1)

Two diodes are described in these three lines. The anode of DBIG10 and the cathode of DBIG20 are connected to node 9. Both use the single DHEFTY model line. DHEFTY has a saturation current of 0.02 pA, a reverse breakdown voltage of 800 volts, and a current at reverse breakdown of 100 mA. However, the diodes are not the same, even though both use the DHEFTY model. Before you give up in confusion, notice the DBIG20 element line which has the number 2.5 in it. 2.5 is the area factor of DBIG20. This means that the physical area of diode DBIG20 will be 2.5 times the area of the DHEFTY diode, and the IS, RS, CJO and IBV parameters of diode DBIG20 will be different from the IS, RS, CJO and IBV parameters of diode DBIG10.

EXAMPLE:

DREF 1 6 TYPE17
.MODEL TYPE17 D(BV = 6.2 IBV = 10M)

The diode named DREF has its anode at node 1 and its cathode at node 6. It appears to be a Zener diode, due to its low reverse breakdown voltage of 6.2 V.

8.2.3 Diode Model Parameters

Notice that if no parameters are specified in the .MODEL MODNAME D model line, then ohmic resistance (RS), transit time

Diode Model Parameters

Name	Parameter	Units	Default
IS	saturation current	A	1.0E-14
N	emission coefficient	-	1
ISR	recombination current parameter	A	0
NR	emission coefficient for ISR	-	2
IKF	high-injection knee current	A	infinite
BV	reverse breakdown voltage	V	infinite
IBV	current at breakdown voltage	A	1.0E-3
NBV	reverse breakdown ideality factor	-	1
IBVL	low-level reverse breakdown knee current	A	1.0E-3
NBVL	low-level rev. breakdown ideality factor	A	1.0E-3
RS	ohmic resistance	Ohm	0
TT	transit-time	sec	0
CJO	zero-bias junction capacitance	F	0
VJ	junction potential	V	1
M	grading coefficient	-	0.5
FC	forward bias depletion capacitance coefficient	-	0.5
EG	activation energy	eV	1.11
XTI	saturation current temperature exponent	-	3.0
TIKF	IKF temperature coefficient (linear)	C^{-1}	0
TBV1	BV temperature coefficient (linear)	C^{-1}	0
TBV2	BV temperature coefficient (quadratic)	C^{-2}	0
TRS1	RS temperature coefficient (linear)	C^{-1}	0
TRS2	RS temperature coefficient (quadratic)	C^{-2}	0
KF	flicker noise coefficient	-	0
AF	flicker noise exponent	-	1

(TT), flicker noise coefficient (KF) and zero-bias junction capacitance (CJO) all default to zero. Also, the default value of the reverse breakdown voltage (BV) is infinity. If BV or IBV are specified in the diode model line, positive values should be used.

8.3 DESCRIBING BIPOLAR JUNCTION TRANSISTORS TO PSpice

8.3.1 BJT Element Line

The general form for an element line for a bipolar junction transistor is

QXXXXXX NC NB NE <NS> MODNAME <AREA> <OFF> <IC=VBE,VCE>

where QXXXXXX is the name of the transistor. NC, NB and NE are the nodes to which the collector, base and emitter are connected, respectively. MODNAME is the model name which is used in an associated .MODEL control line. All of these are required in the BJT element line. The optional parameters are:

NS - the node to which the substrate is connected; if omitted, NS defaults to node 0.

AREA - the area factor which determines how many of the BJT model MODNAME are put in parallel to make one QXXXXXX.

OFF - an initial condition of QXXXXXX for the DC analysis.

IC=VBE,VCE - a specification of initial conditions, for use with the UIC option of the .TRAN control line. This causes PSpice to use VBE and VCE as the initial conditions for base-emitter and collector-emitter voltages, instead of the quiescent operating point junction voltages when a transient analysis is done.

8.3.2   BJT Model Line

The general forms for model lines for bipolar junction transistors are

.MODEL MODNAME NPN<(PAR1=PVAL1 PAR2=PVAL2 ...)>
.MODEL MODNAME PNP<(PAR1=PVAL1 PAR2=PVAL2 ...)>

where MODNAME is the model name given to a BJT in an element line, and NPN or PNP tells PSpice that the semiconductor being modeled is an NPN or PNP BJT and not a diode, FET or MOSFET. PAR is the parameter name, which can be any of those in the parameter list in Section 8.3.3, and PVAL is the value of that parameter. Note that the PAR and PVAL parts of the .MODEL line are optional; the minimum specification of a BJT is simply NPN or PNP.

There are 55 parameters that can be specified in the .MODEL control line for a BJT. The model is adapted from the integral charge control model of Gummel and Poon. If certain parameters are not specified, the model will become the simpler Ebers-Moll model.

EXAMPLE:

QBUFFER 4 10 5 SMALLSIG
.MODEL SMALLSIG NPN(BF = 140)

Bipolar junction transistor QBUFFER has its collector at node 4, its base at node 10 and its emitter at node 5. The substrate of QBUFFER, by default, is node 0. This is an NPN transistor with all the default parameters listed in Section 8.3.3, except that its forward beta is set to 140.

EXAMPLE:

QTOP   3 8 9 HIPOWER 8
QMID   1 2 4 HIPOWER
QBOTTOM6 7 9 HIPOWER 8
.MODEL HIPOWER NPN(BF = 50)

Bipolar junction transistors QTOP and QBOTTOM have area factors of 8, which means their physical areas will be 8 times the area of QMID. All parameters in Section 8.3.3 with an asterisk (*) in the area column will be affected by the area factor. By default, QMID has an area factor of 1.

All three NPN BJTs have a forward beta of 50. Note that only one .MODEL line is needed for these three transistors. Any number of transistors that have the same MODNAME can use the same .MODEL line.

EXAMPLE:

QINPUT 10 11 12 MOD15
.MODEL MOD15 PNP

BJT QINPUT has its collector, base and emitter connected to nodes 10, 11 and 12, respectively. It is a PNP type, with all the default parameters listed in Section 8.3.3.

8.3.3 BJT Model Parameters

The DC model is defined by the parameters IS, BF, NF, ISE, IKF, and NE which determine the forward current gain characteristics, IS, BR, NR, ISC, IKR and NC which determine the reverse current gain characteristics, and VAF and VAR which determine the output conductance for forward and reverse regions. Three ohmic resistances, RB, RC and RE, are included, where RB can be high current dependent. Base charge storage is modeled by forward and reverse transit times TF and TR, the forward transit time TF being bias dependent if desired, and nonlinear depletion layer capacitances which are determined by CJE, VJE and MJE for the B-E junction, CJC, VJC and MJC for the B-C junction, and CJS, VJS and MJS for the C-S (Collector-Substrate) junction. The temperature dependence of the saturation current, IS, is determined by the energy gap, EG, and the saturation current temperature exponent, XTI. Additionally, base current temperature dependence is modeled by the beta temperature exponent XTB in the model.

Name	Parameter	Unit	Default
IS	transport saturation current	A	1.0E-16
BF	ideal maximum forward beta	-	100
NF	forward current emission coefficient	-	1.0
VAF	forward Early voltage	V	infinite
IKF	corner for forward beta high current roll-off	A	infinite
ISE	B-E leakage saturation current	A	01.0E-13
NE	B-E leakage emission coefficient	-	1.5
BR	ideal maximum reverse beta	-	1
NR	reverse current emission coefficient	-	1
VAR	reverse Early voltage	V	infinite
IKR	corner for reverse beta high current roll-off	A	infinite
ISC	B-C leakage saturation current	A	0
NC	B-C leakage emission coefficient	-	21.5
NK	high current roll-off coefficient	-	0.5
ISS	substrate p-n saturation current	A	0
NS	substrate p-n emission coefficient	-	1
RE	emitter resistance	Ohm	0
RB	zero bias base resistance	Ohm	0
RBM	minimum base resistance	Ohm	RB
IRB	current where RB falls halfway to RBM	A	infinite
RC	collector resistance	Ohm	0
CJE	B-E zero-bias depletion capacitance	F	0
VJE	B-E built-in potential	V	0.75
MJE	B-E junction exponential factor	-	0.33
CJC	B-C zero-bias depletion capacitance	F	0
VJC	B-C built-in potential	V	0.75
MJC	B-C junction exponential factor	-	0.33
XCJC	fraction of B-C capacitance connected to base	-	1
CJS	zero-bias collector-substrate capacitance	F	0
VJS	substrate junction built-in potential	V	0.75
MJS	substrate junction exponential factor	-	0
FC	coeff., forward-bias depletion capacitance	-	0.5
TF	ideal forward transit time	sec	0
XTF	coefficient for bias dependence of TF	-	0
VTF	voltage describing VBC dependence of TF	V	infinite
ITF	TF dependency on IC	A	0
PTF	excess phase at freq=1.0/(TF*2PI) Hz	deg	0
TR	ideal reverse transit time	sec	0
QCO	epitaxial region charge factor	Coul.	0
RCO	epitaxial region resistance	Ohm	0
VO	carrier mobility knee voltage	V	10
GAMMA	epitaxial region doping factor	-	1E-11
EG	energy gap for temperature effect on IS	eV	1.11
XTB	forward and reverse beta temp. coeff.	-	0
XTI	temperature exponent for effect on IS	-	3
TRE1	RE temperature coefficient (linear)	C^{-1}	0
TRE2	RE temperature coefficient (quadratic)	C^{-2}	0
TRB1	RB temperature coefficient (linear)	C^{-1}	0
TRB2	RB temperature coefficient (quadratic)	C^{-2}	0
TRM1	RBM temperature coefficient (linear)	C^{-1}	0
TRM2	RBM temperature coefficient (quadratic)	C^{-2}	0
TRC1	RC temperature coefficient (linear)	C^{-1}	0
TRC2	RC temperature coefficient (quadratic)	C^{-2}	0
KF	flicker-noise coefficient	-	0
AF	flicker-noise exponent	-	1

8.4 DESCRIBING JUNCTION FIELD-EFFECT TRANSISTORS TO PSpice

8.4.1 JFET Element Line

The general form for an element line for a junction field-effect transistor is

JXXXXXX ND NG NS MODNAME <AREA> <OFF> <IC=VDS,VGS>

where JXXXXXX is the name of the transistor. ND, NG and NS are the nodes to which the drain, gate and source are connected, respectively. MODNAME is the model name which is used in an associated .MODEL control line. All of these are required in the JFET element line. The optional parameters are:

AREA - the area factor which determines how many of the JFET model MODNAME are put in parallel to make one JXXXXXX.

OFF - an initial condition of JXXXXXX for the DC analysis.

IC=VDS,VGS - a specification of initial conditions, for use with the UIC option of the .TRAN control line. This causes PSpice to use VDS and VGS as the initial conditions for drain-source and gate-source voltages instead of the quiescent operating point junction voltages when a transient analysis is done.

8.4.2 JFET Model Line

The general forms for model lines for junction field-effect transistors are

.MODEL MODNAME NJF<(PAR1=PVAL1 PAR2=PVAL2 ...)>
.MODEL MODNAME PJF<(PAR1=PVAL1 PAR2=PVAL2 ...)>

where MODNAME is the model name given to a JFET in an element line, and NJF or PJF tells PSpice that the semiconductor being modeled is an N-channel or P-channel JFET and not a diode, BJT or MOSFET. PAR is the parameter name, which can be any of those in the parameter list in Section 8.4.3, and PVAL is the value of that parameter. Note that the PAR and PVAL parts of the .MODEL line are optional; the minimum specification of a JFET is simply NJF or PJF.

There are 21 parameters that can be specified in the .MODEL control line for a JFET. The model is derived from the FET model of Shichman and Hodges.

EXAMPLE:

JBUFFER 8 5 7 EASY
.MODEL EASY PJF

JBUFFER has its drain, gate and source connected to nodes 8, 5 and 7, respectively. It is a P-channel JFET, which has the 12 default parameters listed in Section 8.4.3.

EXAMPLE:

```
JTOP      3 8 9   JBIG 5
JMID      1 2 4   JBIG
JBOTTOM 6 7 9   JBIG 5
.MODEL JBIG NJF(VTO = -3.5)
```

Junction field effect transistors JTOP and JBOTTOM have area factors of 5, which means their physical areas will be 5 times the area of JMID. All parameters shown in Section 8.4.3 with an asterisk (*) in the area column will be affected by the area factor. By default, JMID has an area factor of 1.

All three N-channel JFETs have a threshold voltage of -3.5 V. Note that only one .MODEL line is needed for these three transistors. Any number of transistors that have the same MODNAME can use the same .MODEL line.

8.4.3 JFET Model Parameters

The JFET model is derived from the FET model of Shichman and Hodges. The DC characteristics are defined by the parameters VTO and BETA, which determine the variation of drain current with gate voltage, LAMBDA, which determines the output conductance, and IS, the saturation current of the two gate junctions. Two ohmic resistances, RD and RS, are also included. Charge storage is modeled by nonlinear depletion layer capacitances for both gate junctions which vary as the -1/2 power of junction voltage and are defined by the parameters CGS, CGD and PB.

JFET Parameters

Name	Parameter	Unit	Default
VTO	threshold voltage	V	-2.0
BETA	transconductance parameter	A/V**2	1.0E-4
LAMBDA	channel length modulation parameter	1/V	0
IS	gate junction saturation current	A	1.0E-14
N	gate junction emission coefficient	-	1
ISR	gate junction recombination current parameter	A	0
NR	emission coefficient for ISR	-	2
ALPHA	ionization coefficient	1/V	0
VK	ionization knee voltage	V	0
RD	drain ohmic resistance	Ohm	0
RS	source ohmic resistance	Ohm	0
CGD	zero-bias G-D junction capacitance	F	0
CGS	zero-bias G-S junction capacitance	F	0
M	gate junction grading coefficient	-	0.5
PB	gate junction potential	V	1
FC	forward-bias depletion capacitance coeff.	-	0.5
VTOTC	VTO temperature coefficient	V/C	0
BETATCE	BETA exponential temperature coefficient	%/C	0
XTI	IS temperature coefficient	-	3
KF	flicker noise coefficient	-	0
AF	flicker noise exponent	-	1

8.5 DESCRIBING MOS FIELD-EFFECT TRANSISTORS TO PSpice

8.5.1 MOSFET Element Line

The general form for an element line for a metal-oxide semi-conductor field-effect transistor is

MXXXXXX ND NG NS NB MODNAME <L=VAL> <W=VAL>
+ <AD=VAL> <AS=VAL> <PD=VAL> <PS=VAL> <NRD=VAL>
+ <NRS=VAL> <OFF> <IC=VDS,VGS,VBS>

where MXXXXXX is the name of the MOSFET transistor. ND, NG, NS and NB are the nodes to which the drain, gate, source and bulk (or substrate) are connected, respectively. MODNAME is the model name which is used in an associated .MODEL control line. All of these are required in the MOSFET element line. The optional parameters and their meanings are:

L   length of the channel, in meters
W   width of the channel, in meters
AD area of drain diffusion, in square meters
AS area of source diffusion, in square meters

If any of the above parameters are not specified, default values are used. The user may specify the values to be used for these default parameters on the .OPTIONS control line (not to be confused with the .MODEL control line). The parameters on the .OPTIONS control line are DEFL for L, DEFW for W, DEFAD for AD, and DEFAS for AS. Use of defaults done this way simplifies the writing of the input file initially and subsequent file editing that would occur if device geometries are changed. See Appendix A.3 for the default values.

The rest of the optional parameters and their meanings are:

PD   perimeter of drain junction, in meters

PS   perimeter of source junction, in meters

NRD equivalent number of squares of drain diffusion

NRS equivalent number of squares of source diffusion. NRD and NRS values multiply the sheet resistance RSH specified on the .MODEL line for an accurate representation of the parasitic series drain and source resistance of each transistor. PD and PS default to 0.0 while NRD and NRS default to 1.0

OFF indicates an initial condition on the device for DC analysis

IC= sets the initial conditions for drain-source voltage, gate-source voltage and substrate-source voltage. This is intended to be used with the UIC option of the .TRAN control line. This causes PSpice to use the VDS, VGS and VBS values instead of the quiescent operating point voltages when a transient analysis is done

8.5.2 MOSFET Model Line

The general forms for model lines for metal-oxide semiconductor field-effect transistors are

.MODEL MODNAME NMOS<(PAR1=PVAL1 PAR2=PVAL2 ...)>
.MODEL MODNAME PMOS<(PAR1=PVAL1 PAR2=PVAL2 ...)>

where MODNAME is the model name given to a MOSFET in an element line, and NMOS or PMOS tells PSpice that the semiconductor being modeled is an N-channel or P-channel MOSFET and not a diode, BJT or JFET. PAR is the parameter name, which can be any of those in the parameter list in Section 8.5.3, and PVAL is the value of that parameter. Note that the PAR and PVAL parts of the .MODEL line are optional; the minimum specification of a MOSFET is simply NMOS or PMOS.

There are over 50 parameters that can be specified in the .MODEL control line for a MOSFET.

EXAMPLE:

MBUFFER 8 5 7 0 SIMPLE
.MODEL SIMPLE NMOS

Transistor MBUFFER has its drain, gate, source and substrate connected to nodes 8, 5, 7 and 0 respectively. It is an N-channel MOSFET, which has the default parameters listed in Section 8.5.3. It also has geometry information which is either specified in the .OPTIONS control line or will default to those values listed in the .OPTIONS section. Those default values are: DEFL = 100 microns, DEFW = 100 microns, DEFAD = 0 square meters and DEFAS = 0 square meters.

EXAMPLE:

MTOP   3 8 9 5 TYPEA 80U
MMID   1 2 4 5 TYPEA 60U
MBOTTOM   6 7 9 5 TYPEA
.MODEL TYPEA PMOS(KP = 2.5E-5)

Metal-oxide semiconductor field-effect transistors MTOP and MMID have channel lengths of 80 and 60 microns, respectively. MBOTTOM uses the default value in the .OPTIONS control line.

All three P-Channel MOSFETs have a transconductance parameter of 25 microamp/(volt squared). Note that only one .MODEL line is needed for these three transistors. Any number of transistors that have the same MODNAME can use the same .MODEL line.

8.5.3 MOSFET Model Parameters

PSpice provides three MOSFET device models which differ in the formulation of the I-V characteristic. The variable LEVEL specifies the model to be used:

    LEVEL=1 -> Shichman-Hodges
    LEVEL=2 -> MOS2 (as described in [1])
    LEVEL=3 -> MOS3, a semi-empirical model (see [1])
    LEVEL=4 -> BSIM, Berkeley Short-Channel IGFET Model
                (see [2])

MOSFET Parameters

Name	Parameter	Units	Default
LEVEL	model index	-	1
VTO	zero-bias threshold voltage	V	0.0
KP	transconductance parameter	A/V**2	2.0E-5
GAMMA	bulk threshold parameter	V**0.5	0.0
PHI	surface potential	V	0.6
LAMBDA	channel-length modulation (MOS1 and MOS2 only)	1/V	0.0
RD	drain ohmic resistance	Ohm	0.0
RS	source ohmic resistance	Ohm	0.0
RG	gate ohmic resistance	Ohm	0.0
RB	bulk ohmic resistance	Ohm	0.0
RDS	drain-source shunt resistance	Ohm	infinite
CBD	zero-bias B-D junction capacitance	F	0.0
CBS	zero-bias B-S junction capacitance	F	0.0
IS	bulk junction saturation current	A	1.0E-14
PB	bulk junction potential	V	0.8
PBSW	bulk p-n sidewall potential	V	PB
TT	bulk p-n transit time	sec.	0.0
CGSO	gate-source overlap cap. per m. channel width	F/m	0.0
CGDO	gate-drain overlap cap. per m. channel width	F/m	0.0
CGBO	gate-bulk overlap cap. per m. channel length	F/m	0.0
RSH	drain and source diffusion sheet resistance	Ohm/sq.	0.0
CJ	zero-bias bulk junction bottom cap. per sq-meter of junction area	F/m**2	0.0
MJ	bulk junction bottom grading coefficient	-	0.5
CJSW	zero-bias bulk junction sidewall capacitance per meter of junction perimeter	F/m	0.0
JSSW	bulk p-n saturation sidewall current/length	A/m	0.0
N	bulk p-n emission coefficient	-	1
MJSW	bulk junction sidewall grading coefficient	-	0.3
JS	bulk junction saturation current per square-meter of junction area	A/m**2	1.0E-8
TOX	oxide thickness	meter	1.0E-07
NSUB	substrate doping	1/cm**3	0.0
NSS	surface state density	1/cm**2	0.0
NFS	fast surface state density	1/cm**2	0.0
TPG	type of gate material: +1 opp. to substrate -1 same as substrate 0 Al gate	-	1.0
XJ	metallurgical junction depth	meter	0.0
LD	lateral diffusion	meter	0.0
UO	surface mobility	cm**2/V-s	600
UCRIT	critical field for mobility degradation (MOS2 only)	V/cm	1.0E4
UEXP	critical field exponent in mobility degradation (MOS2 only)	-	0.0
UTRA	transverse field coefficient (mobility) (deleted for MOS2)	-	0.0
VMAX	maximum drift velocity of carriers	m/s	0.0
NEFF	total channel charge (fixed and mobile) coefficient (MOS2 only)	-	1.0
XQC	thin-oxide capacitance model flag and coefficient of channel charge share attributed to drain (0-0.5)	-	1.0
KF	flicker noise coefficient	-	0.0
AF	flicker noise exponent	-	1.0
FC	coefficient for forward-bias depletion capacitance formula	-	0.5
DELTA	width effect on threshold voltage (MOS2 and MOS3)	-	0.0
THETA	mobility modulation (MOS3 only)	1/V	0.0
ETA	static feedback (MOS3 only)	-	0.0
KAPPA	saturation field factor (MOS3 only)	-	0.2

Note: For information on using the LEVEL 4 model, refer to the PSpice User's Guide, available from MicroSim Corporation.

The DC characteristics of the MOSFET are defined by the device parameters VTO, KP, LAMBDA, PHI and GAMMA. PSpice computes these parameters if process parameters (NSUB, TOX, . . .) are given, but user-specified values always override. VTO is positive (negative) for enhancement mode and negative (positive) for depletion mode N-channel (P-channel) devices. Charge storage is modeled by three constant capacitors, CGSO, CGDO, and CGBO which represent overlap capacitances, by the nonlinear thin-oxide capacitance which is distributed among the gate, source, drain, and bulk regions, and by the nonlinear depletion-layer capacitances for both substrate junctions divided into bottom and periphery, which vary as the MJ and MJSW power of junction voltage respectively, and are determined by the parameters CBD, CBS, CJ, CJSW, MJ, MJSW and PB. There are two built-in models of the charge storage effects associated with the thin-oxide. The default is the piece-wise linear voltage-dependent capacitance model proposed by Meyer. The second choice is the charge-controlled capacitance model of Ward and Dutton [1]. The XQC model parameter acts as a flag and a coefficient at the same time. As the former it causes the program to use Meyer's model whenever larger than 0.5 or not specified, and the charge-controlled model when between 0 and 0.5. In the latter case its value defines the share of the channel charge associated with the drain terminal in the saturation region. The thin-oxide charge storage effects are treated slightly differently for the LEVEL 1 model. These voltage-dependent capacitances are included only if TOX is specified in the input description, and they are represented using Meyer's formulation.

There is some overlap among the parameters describing the junctions, e.g. the reverse current can be given either as IS (in A) or as JS (in A/m**2). Whereas the first is an absolute value, the second is multiplied by AD and AS to give the reverse current of the drain and source junctions respectively. This methodology has been chosen since there is no sense in always relating junction characteristics with AD and AS entered on the device card; the areas can be defaulted. The same idea also applies to the zero-bias junction capacitances CBD and CBS (in F) on the one hand, and CJ (in F/m**2) on the other. The parasitic drain and source series resistance can be expressed as either RD and RS (in ohms) or RSH (in ohms per square), the latter being multiplied by the number of squares NRD and NRS specified on the device line.

8.6 DESCRIBING GALLIUM ARSENIDE FIELD-EFFECT TRANSISTORS TO PSpice

8.6.1 GaAsFET Element Line

The general form for an element line for a gallium arsenide field-effect transistor is

BXXXXXX ND NG NS MODNAME <AREA>

where BXXXXXX is the name of the GaAsFET transistor. ND, NG and NS are the nodes to which the drain, gate, and source are connected, respectively. MODNAME is the model name which is used in an associated .MODEL control line. All of these are required in the MOSFET element line. The optional parameter and its meaning is

> AREA the area factor which determines how many of the GaAsFET model MODNAME are put in parallel to make one BXXXXXX.

8.6.2 GaAsFET Model Line

The general form for the model line for gallium arsenide field-effect transistors is

.MODEL MODNAME GASFET<(PAR1=PVAL1 PAR2=PVAL2 ...)>

where MODNAME is the model name given to a GaAsFET in an element line, and GASFET tells PSpice that the semiconductor being modeled is a GaAsFET and not a diode, BJT, JFET or MOSFET. PAR is the parameter name, which can be any of those in the GaAsFET parameter list, and PVAL is the value of that parameter. Note that the PAR and PVAL parts of the .MODEL line are optional; the minimum specification of a GaAsFET is simply GASFET.

There are 30 parameters that can be specified in the .MODEL control line for a GaAsFET. The variable LEVEL specifies the model to be used as follows:

> LEVEL=1 -> "Curtice" model (as described in [3])
> LEVEL=2 -> "Raytheon" or "Statz" model (as described in [4]), equivalent to SPICE3 GaAsFET model
> LEVEL=3 -> "TriQuint model (see [5])

EXAMPLE:

B_RFAMP 6 2 4 HOTSTUFF 1.7
.MODEL HOTSTUFF GASFET(level=2 VTO=-2.3 IS=1.5E-14)

Transistor B_RFAMP has its drain, gate and source connected to nodes 6, 2 and 4 respectively. It is a gallium arsenide FET Level 2 model, with an area factor of 1.7, a pinch-off voltage of -2.3 V, and a gate P-N saturation current of 15 fA.

8.6.3 GaAsFET Model Parameters

GaAsFET Parameters

Name	Parameter	Units	Default
LEVEL	model index (1, 2 or 3)	-	1
VTO	pinch-off voltage	V	-2.5
ALPHA	saturation voltage parameter	1/V	2.0
BETA	transconductance coefficient	A/V**2	0.1
B	doping tail extending parameter (level=2 only)	1/V	0.3
LAMBDA	channel-length modulation	1/V	0.0
GAMMA	static feedback parameter (level=3 only)	-	0.0
DELTA	output feedback parameter (level=3 only)	1/(A*V)	0.0
Q	power-law parameter (level=3 only)	-	2
TAU	conduction current delay time	sec	0
RG	gate ohmic resistance	Ohm	0.0
RD	drain ohmic resistance	Ohm	0.0
RS	source ohmic resistance	Ohm	0.0
IS	gate p-n saturation current	A	1.0E-14
N	gate p-n emission coefficient	-	1
M	gate p-n grading coefficient	-	0.5
VBI	gate p-n potential	V	1.0
CGD	zero-bias gate-drain p-n capacitance	F	0.0
CGS	zero-bias gate-source p-n capacitance	F	0.0
CDS	drain-source capacitance	F	0.0
FC	forward-bias depletion capacitance coeff.	-	0.5
EG	bandgap voltage (barrier height)	eV	1.11
XTI	IS temperature exponent	-	0
VTOTC	VTO temperature coefficient	V/deg C	0
BETATCE	BETA exponential temperature coefficient	%/deg C	0
TRG1	RG temperature coefficient (linear)	1/deg C	0
TRD1	RD temperature coefficient (linear)	1/deg C	0
TRS1	RS temperature coefficient (linear)	1/deg C	0
KF	flicker noise coefficient	-	0.0
AF	flicker noise exponent	-	1.0

[1] A. Vladimirescu and S. Liu, "The Simulation of MOS Integrated Circuits Using SPICE2," ERL Memo No. ERL M80/7, Electronics Research Laboratory, University of California, Berkeley, Oct. 1980.

[2] B.J. Sheu, D.L. Scharfetter and P.K. Ko, "SPICE2 Implementation of BSIM," ERL Memo No. ERL M85/42, Electronics Research Laboratory, University of California, Berkeley, May 1985.

[3] W.R. Curtice, "A MESFET model for use in the design of GaAs integrated circuits," *IEEE Transactions on Microwave Theory and Techniques*, MTT-32, pp. 471-473 (1984).

[4] H. Statz, P. Newman, I.W. Smith, R.A. Pucel and H.A. Haus, "GaAsMESFET computer model for SPICE," *IEEE Transactions on Electron Devices*, ED-34, pp. 160-169 (1987).

[5] A.J. McCamant, G.D. McCormack and D.H. Smith, "An Improved GaAs MESFET Model for SPICE," *IEEE Transactions on Microwave Theory and Techniques*, MTT-38, no. 6, pp. 822-824, (1990).

CHAPTER SUMMARY

PSpice has built-in models for five kinds of semiconductors: junction diode, bipolar junction transistor, junction field-effect transistor, MOS field-effect transistor, and gallium arsenide field-effect transistor.

In order to describe a semiconductor in an input file, both an element line and a .MODEL line are needed. However, many devices of the same type with identical parameters (e.g. NPN BJT with BF = 125) can use the same .MODEL line.

For many circuit simulations only a few key parameters need to be specified in the .MODEL line to PSpice. The default parameters give excellent results in a large proportion of analyses.

Chapter 9

SUBCIRCUITS

9.1 THE NEED FOR SUBCIRCUITS IN PSpice INPUT FILES

Many circuits are made with basic electronic building blocks such as operational amplifiers, logic gates and comparators. For example, an active filter might contain six identical op-amps. One way to write an input file for such an active filter for PSpice analysis would be to describe the op-amp six times. If it took 10 element lines to describe the op-amp, then for the active filter 60 element lines would be needed just for the op-amps. In addition to 60 element lines for op-amps, additional element lines would be needed for resistors and capacitors external to the op-amps. Using this approach is tedious work; fortunately, PSpice permits us to define an often-used circuit block only once in each input file. Then that circuit block can be used many times by writing a single element line each time the circuit block is used. PSpice calls such circuit blocks subcircuits.

9.2 HOW TO USE SUBCIRCUITS

9.2.1 Control Lines Used For Subcircuit Definition

In order to use a subcircuit in an input file, the subcircuit must be defined. The format of the first line of a subcircuit definition is

.SUBCKT SUBNAME N1 <N2 N3 ... >

and the last line of a subcircuit definition is

.ENDS <SUBNAME>

The .ENDS control line should not be confused with the last control line of the entire input file, .END. In a PSpice input file with three subcircuits, there would be three .ENDS SUBNAME control lines and only one .END control line.

SUBNAME is the name you will call the subcircuit. N1, N2 . . . are those nodes of the subcircuit that will be connected to other nodes in the circuit; they are called the external nodes of the subcircuit. Node zero may not be an external node of the subcircuit. In addition to the external nodes, a subcircuit can have any number of internal nodes. Internal nodes are nodes within the subcircuit definition which are not external nodes; node zero may be an internal node. Such internal nodes are local to that subcircuit (except for node zero), and are not considered by PSpice to be connected to the circuit outside of the subcircuit, <u>even if the node numbers are the same</u>. However, in the interests of clarity, not to mention kindness to others who may look at your input file, it is best not to use the same node numbers within a subcircuit and also in the circuit.

<u>Allowed</u> within a subcircuit, between the .SUBCKT and .ENDS control lines, are element lines of the subcircuit, semiconductor device models, other subcircuit definitions, and even subcircuit calls. <u>Not allowed</u> within a subcircuit definition are control lines.

The .ENDS line must have the optional SUBNAME included only if subcircuit definitions are nested (only if a subcircuit definition contains another subcircuit definition).

9.2.2   Element Line to Use a Subcircuit in a Circuit

XYYYYYY N1 <N2 N3 ... > SUBNAME

XYYYYYY is the element name of the pseudo-element which is defined elsewhere in the input file as SUBNAME.   N1 . . .   are
the circuit nodes to which the subcircuit is connected.

9.2.3   Example of Subcircuit Use: An R-C Circuit

Consider the R-C circuit fed by a sinusoidal voltage source
shown in Fig. 9.1.   First we'll do an AC analysis without subcircuits.   The PSpice input file shown in Fig. 9.2 contains 13

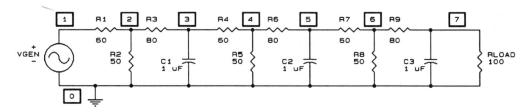

**Figure 9.1 R-C Circuit with Sinusoidal
Voltage Source.**

```
RC1.CIR    LONG R-C NETWORK
VGEN     1   0   AC    1
R1       1   2   60
R2       2   0   50
R3       2   3   80
C1       3   0   1U
R4       3   4   60
R5       4   0   50
R6       4   5   80
C2       5   0   1U
R7       5   6   60
R8       6   0   50
R9       6   7   80
C3       7   0   1U
RLOAD    7   0   100
.AC  DEC  25   100   1MEG
.PROBE
.OPTIONS NOPAGE
.END
```

**Figure 9.2 Input File RC1.CIR for R-C
Circuit.**

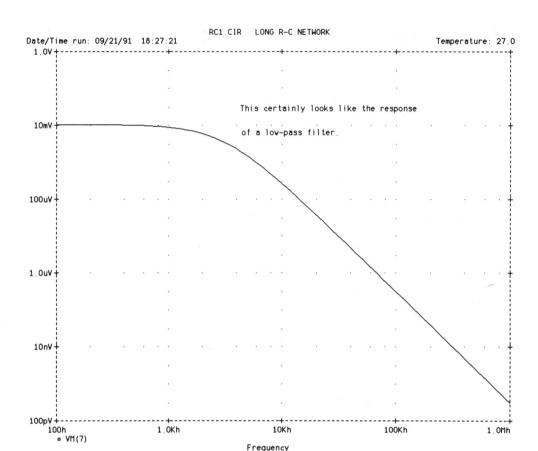

Figure 9.3   Bode Plot of R-C Circuit.

element lines.   Figure 9.3 is the PSpice output file, containing a graph of load voltage versus frequency.   As we might predict, the network is a low-pass filter, where the load voltage
rolls off as frequency increases.

By examining the circuit, we can see that the circuit combination of three resistors (60, 50 and 80 ohms) and one capacitor is repeated three times.   These four components can be
defined as a subcircuit, shown with the subcircuit definition
in Fig. 9.4.   Figure 9.5 is the same schematic diagram as Fig.
9.1, with the subcircuits illustrated.   The input file shown
in Fig. 9.6 requires only nine element lines (compared with 13
in Fig. 9.2) to describe the circuit.   The analysis results,
shown in Fig. 9.7, are the same.

SUBCIRCUIT DEFINITION

```
.SUBCKT   TEE   1  2
RLEFT     1  8  60
RRIGHT    8  2  80
RMID      8  0  50
CRIGHT    2  0  1U
.ENDS   TEE
```

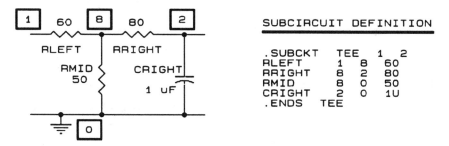

Figure 9.4 Schematic and Definition of Subcircuit.

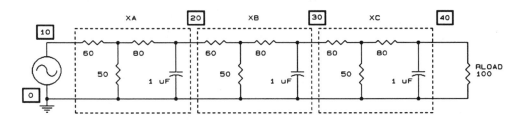

Figure 9.5 R-C Circuit with Subcircuits Shown.

```
RC2.CIR   LONG R-C DONE WITH SUBCIRCUITS
VGEN      10   0    AC    1
.SUBCKT    TEE  1  2
RLEFT     1  8   60
RMID      8   0   50
RRIGHT    8   2   80
CRIGHT    2   0   1U
.ENDS  TEE
XA        10  20    TEE
XB        20  30    TEE
XC        30  40    TEE
RLOAD     40  0    100
.AC  DEC  25  100  1MEG
.PROBE
.OPTIONS  NOPAGE
.END
```

Figure 9.6 Input File RC2.CIR for R-C Circuit with Subcircuits.

Of course, if you were modeling a flip-flop using many nand gates, the savings in element lines by using subcircuits for the nand gates would be even more noticeable.

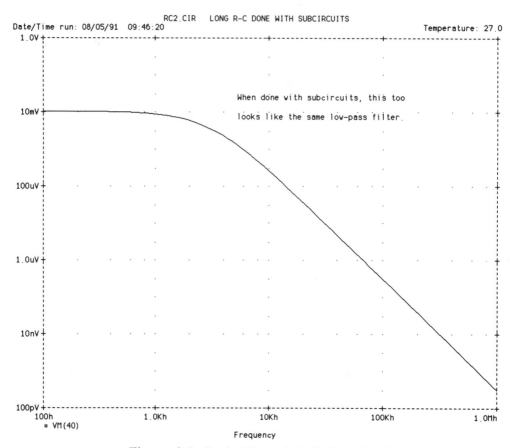

When done with subcircuits, this too
looks like the same low-pass filter.

□ VM(40)

Frequency

**Figure 9.7 Bode Plot of R-C Circuit with
Subcircuits.**

CHAPTER SUMMARY

Subcircuits can simplify and dramatically shorten input files
which use the same circuitry many times.

Subcircuits may be nested within other subcircuits. For example, a subcircuit for a shift register could have a flip-flop
subcircuit within its definition, and the flip-flop subcircuit
could have a nand gate subcircuit among its subcircuit element
lines.

Try to use node numbers within subcircuit definitions which
are different from those used outside the subcircuit. Though
this is not required, doing so avoids confusion when another
person looks at the input file.

Chapter 10

TRANSMISSION LINES

10.1 TRANSMISSION LINES ARE LOSSLESS ?

Yes, if you are using PSpice, the basic transmission line element is considered to be lossless. This is quite reasonable for a circuit analysis program with integrated circuit emphasis, since transmission lines used within an integrated circuit (likely to be stripline or microstrip) are going to be very short, both physically and electrically. The loss in a short transmission line is normally quite small (a fraction of a dB) and therefore negligible for most purposes.

If you wish to add loss to a PSpice transmission line, this can be done by using lumped (as opposed to distributed) R or G elements. The dominant loss mechanism in most practical transmission lines is the conductor loss due to skin effect, not the dielectric loss. You can simulate a long and lossy line by breaking it up into shorter sections and adding a small resistance in series with each short section to represent the distributed conductor loss. Although it is tempting to use extremely short sections (each with a very small series R for loss), there is a tradeoff involved with this. In transient analysis, PSpice will make the timestep (or internal computing interval) less than or equal to one-half the minimum transmission delay of the shortest transmission line. This can make a PSpice transient analysis take a very long time if short transmission lines are used.

Another problem is that transmission line loss increases with frequency. It is not possible to specify a frequency-dependent resistor, so line loss simulated by lumped resistors will be valid only at one frequency. The PSpice limitation that transmission lines are lossless is not a serious one. In the examples that follow, you will see that PSpice can be used to solve an assortment of transmission line problems which would be all but impossible to solve in a reasonable time without a computer.

10.2 TRANSMISSION LINE ELEMENT LINE

The two formats for an element line describing a transmission line are

    TXXXXXX NA+ NA- NB+ NB- Z0=ZVAL F=FREQ
        + <NL=NLENGTH> <IC = VA, IA, VB, IB>

or

    TYYYYYY NA+ NA- NB+ NB- Z0=ZVAL TD=TVALUE
        + <IC = VA, IA, VB, IB>

TXXXXXX is the element name of the transmission line. NA+ and NA- are, respectively, the positive and negative nodes of the A side of the transmission line. NB+ and NB- are, respectively, the positive and negative nodes of the B side of the transmission line. Z0 is the characteristic impedance of the transmission line, in ohms. (The 0 in Z0 is the number 0, not the capital letter O.)

PSpice must be told the length of a transmission line. There are two ways to do this; either one is acceptable, and you can easily convert from one to the other. The two ways are:

1. Specify frequency and electrical length: FREQ is the frequency at which the transmission line is NLENGTH wavelengths long. The NLENGTH parameter is dimensionless, since it is the normalized electrical length of the line. NLENGTH is the physical length of the line (in m) divided by the wavelength in the line (in m). The wavelength in the line is the free-space wavelength multiplied by the velocity factor of the line. If the optional NL = NLENGTH

parameter is omitted, PSpice assumes a value of .25. That is, if the length of a transmission line is not specified, PSpice will consider it to be a one-quarter wavelength section.

2. Specify the transmission delay: TD is the transmission delay of the transmission line, measured in seconds.

With each of the methods for specifying the electrical length of a transmission line you can optionally specify initial conditions at each end of the line. VA and IA are the voltage across and current into the A side of the line, while VB and IB are the voltage across and current into the B side of the line. A positive voltage means that the V+ node is more positive than the V- node on that side, while a positive current means that current is flowing into the V+ node and out of the V- node. It is important to note that these initial conditions will have an effect only if the .TRAN control line contains the option UIC (use initial conditions).

A single transmission line element line models only one propagating mode. In those circuits where the four nodes are distinct, two modes may occur: center conductor to shield and shield to ground. In order to model such a situation, two transmission line element lines should be used.

10.3 EQUIVALENCY BETWEEN TRANSMISSION DELAY AND ELECTRICAL LENGTH

Since there are two methods for telling PSpice the length of a transmission line, it may be of use to look at how each method works and how to convert from one method to the other.

In general, we can say that the transmission delay, TD, of a transmission line can be found from

TD = physical length/velocity

where velocity is the phase velocity in the line, given by

velocity = (velocity factor) free-space velocity

The velocity factor is less than or equal to 1.00. Also, the normalized electrical length, NL, is

NL = physical length/wavelength

where wavelength is the wavelength in the line, determined by

wavelength = velocity/frequency

Normalized electrical length can then be expressed as

NL = physical length (frequency)/velocity

Since physical length = TD(velocity) = NL(velocity)/frequency then

TD = NL/frequency, and NL = frequency(TD)

EXAMPLES:

1. A transmission line with a characteristic impedance of 50 ohms is 12 m long, with a velocity factor of 0.66. a. What is the transmission delay? b. What is the normalized electrical length?

   a. velocity = velocity factor (c), and c = 3E8 m/s, so velocity = 0.66 (3E8 m/s) = 2E8 m/s.
      TD = length/velocity = 12 m/(2E8 m/s) = 60 ns.

   b. Before the normalized electrical length can be determined, a frequency must be specified. Let's use 30 MHz. NL = frequency(length)/velocity = 30E6(12)/2E8, so NL = 1.8 wavelengths.

2. Write the element line for the above transmission line two different ways. Call it TSTUB.

   TSTUB 1 2 3 4 Z0 = 50 TD = 60N

   TSTUB 1 2 3 4 Z0 = 50 F = 30MEG NL = 1.8

The element lines above assume TSTUB is connected to nodes 1 and 2 at the A side, and to nodes 3 and 4 at the B side.

10.4   SAMPLE TRANSMISSION LINE PROBLEMS

10.4.1 Single Pulse into Transmission Line

This problem will show how PSpice can be used in the time domain to look at a single short pulse at the sending end and the receiving end of a physically short transmission line.

Figure 10.1 shows a pulse voltage source, with 50-ohm output impedance, connected to a 50-ohm transmission line which is terminated in a 50-ohm matched load. The PSpice input file TLINE1.CIR, Fig. 10.2, specifies that the pulse goes from 0 V to 10 V after a 1 ns delay from time-zero, and has a pulse width of 3 ns.

Transmission line TSHORT has a transmission delay of 2 ns. We should therefore expect that whatever voltage changes appear at the input to the transmission line (node 2) should occur 2 ns later at the output (node 3). The results of the PSpice transient analysis, displayed by Probe, are shown in Fig. 10.3. Voltage V(2) is shown in the upper plot, and the lower plot is V(3).

The input voltage, V(2), rises from 0 V to 5 V at 1 ns and remains 5 V for 3 ns. The output voltage, V(3), does the same as V(2) except that it is delayed by 2 ns. Since the transmission line is terminated in a matched load, no reflections are seen. If the load is not matched, a reflection will occur at the mismatch, and the input voltage will change at a time equal to twice the transmission delay of the line. This is the basis of time-domain reflectometry (TDR) which is used at the input side to examine transmission lines for defects.

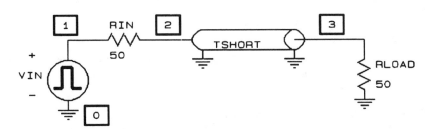

Figure 10.1 Pulse into Transmission Line.

```
TLINE1.CIR  PULSE INTO TRANSMISSION LINE
*      TO ILLUSTRATE THE TIME DELAY OF A LINE
VIN  1  0  PULSE(0 10 1N 0 0 3N)
RIN  1  2  50
TSHORT 2  0  3  0  Z0=50  TD = 2N
RLOAD 3  0  50
.TRAN .02N  7N  0  .01N
.PROBE
.END
```

Figure 10.2 Input File TLINE1.CIR.

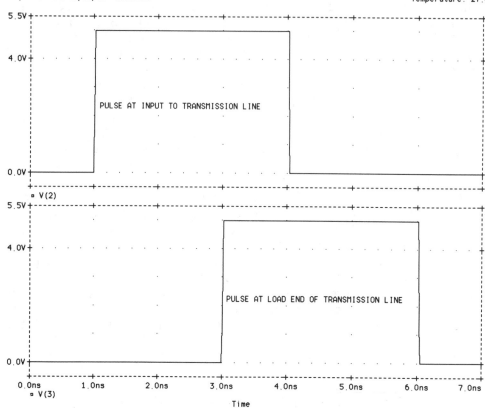

**Figure 10.3 Pulse at Input and Load End
of Transmission Line.**

PSpice can be used to predict in advance what a TDR display
should be for a certain transmission line circuit.

Section 10.4.5 is an example of time-domain reflectometry
with mismatches between both generator/transmission line and
transmission line/load.

10.4.2 Transmission Line Balun Power Bandwidth

PSpice will be used to determine the power bandwidth of a
transmission line balanced-unbalanced (balun) transformer.
Baluns are used to connect coaxial transmission lines to bal-
anced transmission lines or antennas, and utilize a one-half
wavelength section of transmission line to accomplish the
impedance transformation.

It is a very simple task to analyze how baluns work at the design frequency, where the line is indeed one-half wavelength long. However, it is not at all simple to analyze how the balun will perform at frequencies above or below the design frequency. PSpice can be used to great advantage to perform a frequency sweep of the input voltage, and to plot the output voltage versus frequency. We can tell the power bandwidth of the balun circuit by observing when the output voltage magnitude drops to 0.707 of its value (a drop of 3 dB) at the design frequency.

The circuit is shown in Fig. 10.4. The 75-ohm coaxial transmission line impedance is converted by the balun to the 300-ohm impedance of the load. The design frequency of the balun is 100 MHz, so the transmission line T2 is 0.5 wavelength long at 100 MHz. Figure 10.5 shows the input file, BALUN.CIR, in which the AC analysis control line sweeps the frequency linearly from 20 MHz to 180 MHz.

The Probe display of the PSpice analysis output data (Fig. 10.6) shows the graph of the voltage magnitude across the load resistor, VM(30,40), to be a maximum of 20 V at 100 MHz. Using the 0.707 criterion, the cursors in Probe show that the voltage falls to 0.707(20 V), or 14.1 V at 47.6 MHz and 152.4

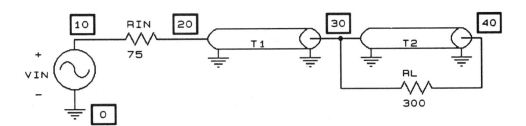

Figure 10.4 Transmission Line Balun Circuit.

```
BALUN.CIR    TRANS. LINE BALUN
VIN      10  0  AC  20
RIN  10  20   75
T1   20  0  30  0   Z0=75  F=100MEG NL=1.0
T2   30  0  40  0   Z0=75  F=100MEG NL=0.5
RL   30  40  300
.AC LIN 401  20MEG 180MEG
.PROBE
.END
```

Figure 10.5 Input File BALUN.CIR.

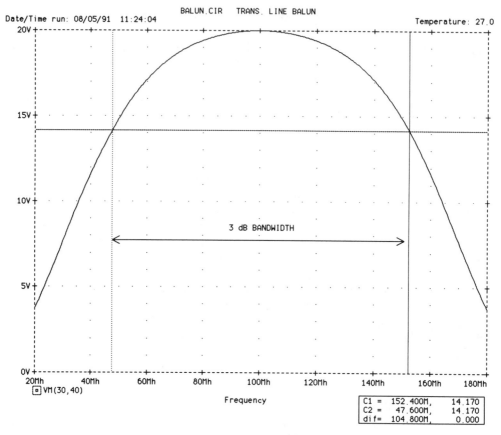

3 dB BANDWIDTH

C1 =	152.400M,	14.170
C2 =	47.600M,	14.170
dif=	104.800M,	0.000

Figure 10.6 Graph of Load Voltage vs. Frequency.

MHz. Thus, PSpice has directly provided the information to determine the power bandwidth to be about 152.4 - 47.6, or 104.8 MHz. The power bandwidth is not necessarily the same as the usable bandwidth, since the voltage standing-wave ratio, or VSWR, may be acceptable only over a much narrower bandwidth. The higher the VSWR, the more power is reflected by the load back to the source. The next problem will show how to make PSpice provide information which can be used to calculate VSWR at each frequency.

10.4.3 Transmission Line Balun Input Impedance

A very simple modification can be made to a circuit to make PSpice calculate the input impedance of a circuit. In this problem we will slightly change the circuit of the previous problem to allow Probe to graph the input impedance of the balun circuit at each frequency. From this input impedance data the VSWR at each frequency can easily be calculated, and a VSWR-based bandwidth can be determined.

If an AC current source of 1 A magnitude and 0 degrees phase is connected to the input of a circuit, then the phasor voltage at the input is the product of current and impedance:

$$V = I\ (Z) = 1.0\ (Z) = Z$$

Thus, by replacing the AC voltage source and 75-ohm resistor in the previous problem with a 1 A current source, PSpice will generate input impedance data directly. Refer to the modified schematic diagram in Fig. 10.7 which shows current source IIN. Figure 10.8 shows the input file, BALUN2.CIR, for this circuit. Since the phase is not specified in the IIN element line, the default value of 0 degrees will apply. Notice that the + node for IIN is 0, and the - node is 20. That means that positive current flows up from ground (node 0) through IIN and into node 20.

The Probe display shown in Fig. 10.9 has two plots which show the complex input impedance in polar form: VM(20) and

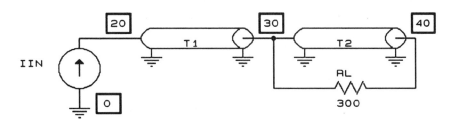

Figure 10.7 Transmission Line Balun Circuit, Current Input.

```
BALUN2.CIR    TRANS. LINE BALUN, 1A CURRENT INPUT
IIN      0  20  AC  1
*        Using 1A in, Vin = Zin
T1    20  0   30  0   Z0=75  F=100MEG  NL=1.0
T2    30  0   40  0   Z0=75  F=100MEG  NL=0.5
RL    30  40  300
.AC LIN 401   60MEG  140MEG
.PROBE
.END
```

Figure 10.8 Input File BALUN2.CIR.

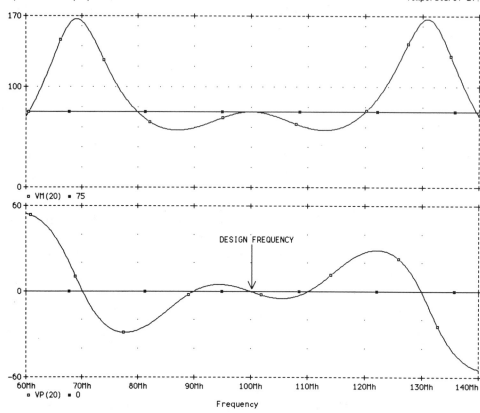

**Figure 10.9 Input Impedance of Balun
Circuit in Polar Form.**

VP(20).   These are the magnitude and phase, respectively, of
ZIN.   If a table of input impedances was needed a .PRINT control line could be added to the PSpice input file, and the
table would be put in the output file.

The input impedance data, when converted into VSWR data,
indicates that the VSWR at 60 MHz and 140 MHz exceeds 3.   The
conversion method can be found in many texts on electronic com-
munications with a chapter on transmission lines.   While not
done in this analysis, the data at 48 MHz and 152 MHz indicate
that the VSWR at those frequencies is 5.66.   A VSWR that high
would be unacceptable for most applications.

10.4.4 Impedance Match with Quarter Wavelength Transmission Line Transformer

One method of matching a load to a transmitter is to use a quarter-wavelength section of transmission line to transform the impedance at its load side to the characteristic impedance of the system. Refer to Fig. 10.10. The purpose of T3 (a 50-ohm line) is to transform the very low resistance of the load, 5 ohms, to 500 ohms at node 4. Transmission line T2 (not a 50-ohm line) is a quarter-wavelength long and transforms the impedance of 500 ohms at node 4 to 50 ohms at node 3, thus creating a perfect match at the design frequency.

The input file in Fig. 10.11, MATCH.CIR, has element lines to describe the circuit, and calls for an AC analysis. The results of the AC analysis will be plotted by Probe. The voltage magnitude across the load, V(5), is plotted two ways in Fig. 10.12. The lower plot is the magnitude of node 5

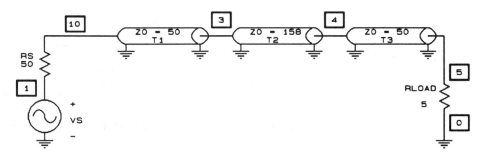

Figure 10.10 Quarterwave Transformer Transmission Line Match Circuit.

```
MATCH.CIR    0.25 WAVELENGTH TRANS. LINE MATCH
VS       1   0  AC  141.42
RS       1  10  50
T1      10   0  3  0  Z0=50  F=50MEG  NL=2
T2       3   0  4  0  Z0=158.11  F=50MEG  NL=0.25
*        SQRT(500*50) = 158.11, 0.25 WAVE MATCH
T3       4   0  5  0  Z0=50  F=50MEG  NL=0.25
*        T3 SECTION TRANSFORMS 5 OHM RLOAD TO 500 OHM
RLOAD    5   0  5
.AC LIN 401  10MEG  90MEG
.PROBE
.END
```

Figure 10.11 Input File MATCH.CIR.

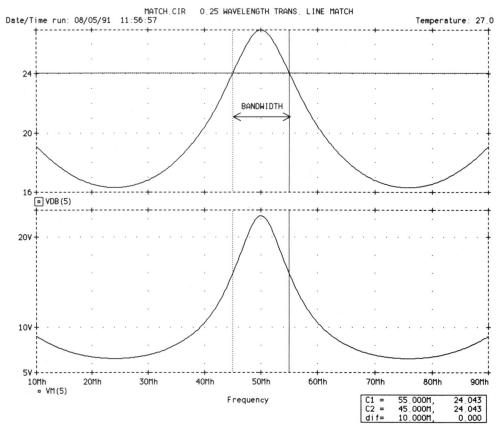

Figure 10.12 Graphs of Load Voltage vs. Frequency.

voltage, while the upper plot shows 20*LOG(node 5 voltage). The decibel data in the upper plot makes it easy to evaluate the half power, or -3 dB, bandwidth of this matching circuit.

The VDB(5) graph shows that load voltage is a maximum of 27 dBV at the design frequency of 50 MHz, and falls off on either side. The 3 dB bandwidth is the range of frequencies over which the magnitude is within 3 dB of 27 dBV. The cursors in Probe easily show that the bandwidth is from 45 MHz to 55 MHz, a 10 MHz range.

10.4.5 Time-Domain Reflectometry With Mismatched Input and Output

An excellent way to understand how pulses behave on transmission lines is to use a time-domain reflectometer (TDR), which sends a train of voltage steps into the sending end of a trans-mission line, and displays the reflections when they arrive back at the sending end. By observing the sending-end volt-age, with a knowledge of voltage reflection coefficients you can quickly determine both the length of the transmission line and the nature of the load impedance. A similar process is used by bats (in air) and by porpoises (in water); it is called SONAR.

Ordinarily, the TDR is matched to the transmission line (i.e., a 50-ohm TDR is used on 50-ohm lines). This is desira-ble because the reflected pulses arrive at the sending end, are absorbed by the generator impedance, and do not get re-flected back towards the load again. An excellent way to rea-lize how little you really know about pulses on transmission lines is to use a TDR which is not matched to the line. In practice this is avoided, as the backward-traveling pulse re-flected from the load and forward-traveling pulse reflected from the generator coincide at the input and make interpreta-tion of the display quite difficult.

This example illustrates a pulsed TDR, with a 200-ohm out-put impedance, connected to a 50-ohm line. As if that's not bad enough, the load on the line is a short circuit. When the voltage at the input is observed, confusion reigns. In order to make what is happening understandable, we can use PSpice to peek at the voltage in the middle of the transmission line.

Figure 10.13 shows the pulse generator with 200-ohm output impedance connected to two 50-ohm transmission lines in ser-ies. Each line has a time delay of 0.25 ns, so the total one-way transmission time is 0.5 ns. Using two lines in series allows us to view the voltage at the midpoint, where the forward-traveling and reverse-traveling pulses do not coin-cide. In the PSpice input file, Fig. 10.14, the load resistor is listed as 1E-6 ohm, as PSpice does not allow zero-ohm resis-tors.

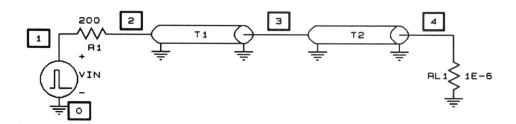

Figure 10.13 Time-Domain Reflectometry Circuit.

```
GNIPGNOP.CIR      Bouncing Pulses on a Trans. Line
VIN   1   0   PULSE(0   2   1N   1P   1P   .1N)
RIN   1   2   200
T1    2   0   3   0   Z0 = 50   TD = 0.25N
T2    3   0   4   0   Z0 = 50   TD = 0.25N
*         The load R below is nearly a short.
RL1   4   0   1U
.TRAN   0.03N   5.5N   0   0.01N
.PROBE
.END
```

Figure 10.14   Input File GNIPGNOP.CIR.

The Probe display of analysis results, Fig. 10.15, shows
the voltage at the generator, V(2), on the upper plot and the
midpoint voltage, V(3), on the lower plot.   The LABEL function
of Probe has been used to mark each pulse with its direction
of travel: F (forward) or R (reverse).   On the upper plot, the

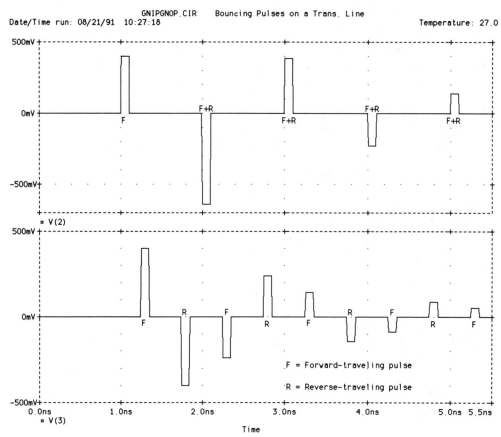

**Figure 10.15   Graphs of Pulses at Input
and Midpoint of Transmission Line.**

first pulse, of 400 mV, is seen at 1 ns. 400 mV is easily explained as the voltage division of the 2 V generator pulse between the 200-ohm generator resistance and the 50-ohm line impedance. However, at 2 ns a pulse is seen, whose amplitude of -640 mV may not be so obvious; it is the sum of the -400 mV pulse reflected back from the load and the -240 mV pulse reflected back towards the load by the generator.

The voltage reflection coefficient of the load is -1.0, while the generator has a voltage reflection coefficient of +0.6. Observing the lower plot of the voltage at the transmission line midpoint, you can see much more clearly what is happening. Every time a forward-traveling pulse (labeled F) is reflected at the load, it comes back with the opposite polarity (multiplied by -1); every time a reverse-traveling pulse (labeled R) is reflected at the generator, it comes back with the same polarity but multiplied by +0.6.

PSpice has been used here to give insight into the behavior of a transmission line circuit, which would be difficult to obtain in the laboratory.

CHAPTER SUMMARY

PSpice recognizes lossless transmission lines only. It is possible to add pseudo-distributed losses to a length of transmission line by breaking it up into many short lengths with series resistance added. However, this technique may lead to substantially increased transient analysis computation time.

The element line for any transmission line must include the characteristic impedance, Z0, and an indication of the line length. The two ways of indicating length are transmission delay (in seconds) or normalized electrical length (in wavelengths) and frequency (in Hz).

Time-domain reflectometry problems (using transient analysis) and steady-state impedance and bandwidth solutions (using AC analysis) can be done easily with PSpice for transmission line circuits.

Chapter 11

HOW TO CHANGE
SEMICONDUCTOR MODELS

11.1   WHY CHANGE SEMICONDUCTOR MODELS?

As you use PSpice to model semiconductor circuits, you will discover that the built-in semiconductor models for diode, BJT, JFET, MOSFET and GaAsFET, with default parameters, do not always match the device characteristics of the semiconductors you are using. For example, you may have a circuit containing a superbeta bipolar transistor with a forward current gain, or beta, of 400. The default beta in a PSpice BJT is 100. Obviously the PSpice BJT model would have to be changed to simulate the circuit accurately. This is easily done.

Another instance in which you may want to change the parameters of a semiconductor model is when a circuit may be built with off-the-shelf components. Beta of BJTs and transconductance of FETs can vary widely from what manufacturers call "typical" values. Perhaps you designed the circuit assuming a typical beta of 100; the beta values of a production run of discrete transistors might vary from 40 to 250. In order to see how the modeled circuit performs with those extremes of beta, it is necessary for you to change the PSpice BJT model.

This chapter will provide an introduction to changing parameters for the five semiconductor devices PSpice recognizes.

For the reader who wishes to dig deeper into this topic, there are some excellent references listed in the bibliography in Appendix D. MicroSim Corporation, supplier of PSpice circuit analysis software, also offers Parts, a program which can generate very accurate models for all types of semiconductors when given manufacturer's data sheet information or measurement data from the laboratory. Parts actually creates the .MODEL line for the semiconductor, which can then be added to the circuit input file. The evaluation version of Parts will only work for diodes.

Another way to obtain accurate models for semiconductors is to contact the manufacturers. Often, they provide very extensive SPICE models for the devices they sell, in order to encourage designers to use their semiconductors. Some even have floppy disks available with models for the devices in library files. The evaluation version of PSpice comes with a limited library of semiconductors and other devices. The PSpice production version library includes models for over 3,500 components.

11.2 CHANGING THE PSpice DIODE MODEL

The table of diode model parameters in Section 8.2.3 lists 25 parameters. The DC characteristics are determined by IS (the saturation current) and N (the emission coefficient). Charge storage effects are specified by TT (transit time) and a nonlinear depletion layer capacitance which is determined by CJO (zero-bias junction capacitance), VJ (junction potential) and M (grading coefficient). The temperature dependence of the saturation current is determined by EG (activation energy) and XTI (saturation current temperature exponent). Reverse breakdown characteristics depend upon BV (reverse breakdown voltage) and IBV (current at breakdown voltage). Both BV and IBV should be expressed as positive numbers.

Yet another parameter which affects IS, RS, CJO and IBV is the area factor (AREA), which is listed not in the .MODEL line but rather in the DXXXXXX element line (AREA is optional). The area factor determines how many of the diode model MODNAME are put in parallel to make one DXXXXXX.

11.2.1 The DC Diode Model

Figure 11.1 shows the DC model for a junction diode. Note that the voltage VD is the junction voltage and does not in-

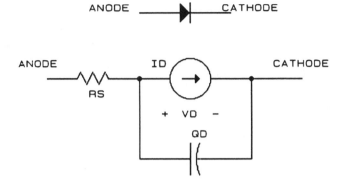

Figure 11.1 DC Model for a Diode.

clude the ohmic voltage drop across the resistance RS. The equation for diode current ID is

$$ID = IS [exp(VD/N*VT) - 1]$$

where IS is the saturation current, N is the emission coefficient (normally set to 1.0), and $VT = k*T/q$. k is Boltzmann's constant, 1.38E-23 Joule/Kelvin, T is the absolute temperature in Kelvin, and q is the charge on an electron, 1.6E-19 C. At a temperature of 25 deg C, $k*T/q = VT = 26$ mV.

Solving the ID equation above for VD gives

$$VD = N * VT * ln(ID/IS +1)$$

The two ways to change the forward voltage drop of the diode for a given forward current are to change N and to change IS. Minor modification can be made to IS based on I-V measurements in the lab. For example, the default value of IS is 1E-14 A. This is a typical value for a silicon integrated circuit diode. If you wanted to model a discrete germanium diode, a typical value of IS would be 5E-9 A. It should be noted that IS also affects the reverse leakage current.

EXAMPLE: How can the IS parameter be obtained?

Perhaps the simplest way is to obtain data which give the diode voltage and current at one point. For example, if the forward voltage across a germanium diode was 316 mV when the diode current was 1 mA, the equation for ID above can be rearranged to give

$$IS = ID/[exp(VD/NVT) - 1]$$

Substituting the ID and VD values, and using a value of 1.0 for N and 26 mV for VT (the room temperature value), we obtain

$$IS = 1E-3/[\exp(0.316/(1.0*26E-3))-1] = 5E-9 = 5 \text{ nA}$$

The use of the evaluation version of Parts, a PSpice option, for obtaining models for diodes will be covered in Section 11.4.

11.3 CHANGING PSpice TRANSISTOR MODELS

The table of bipolar transistor model parameters in Section 8.3.3 lists 55 parameters. Junction field-effect transistors are characterized by 21 parameters, shown in Section 8.4.3. Section 8.5.3 lists 49 parameters for MOSFETs. A substantial discussion of transistor parameters for BJT, JFET and MOSFET transistors, and the methods used for creating accurate models of manufactured transistors, is beyond the scope of this book. There are some excellent references on this topic listed in Appendix D.

For many circuits, in order to simulate circuit behavior accurately when a PSpice analysis is performed, the only transistor parameters that need to be changed are primary gain parameters. These would be ideal maximum forward beta (BETA) for a BJT, transconductance parameter (BETA) for a JFET, and transconductance parameter (KP) for a MOSFET. It is desirable to specify the minimum number of parameters for a semiconductor device which will provide a reasonably accurate model for two reasons:

1. Obtaining many of the parameters is often very difficult and time consuming; manufacturers' data sheets and laboratory tests of transistors do not directly provide the parameters which PSpice uses. An example for a BJT is the Early voltage, both forward and reverse. These two parameters (VAF and VAR) usually must be calculated from other information (h parameters, or characteristic curves which result from laboratory tests).

2. Computation time goes up as the number of parameters specified for each semiconductor device increases. Circuit designers generally work with simple models in order to save time, and often do achieve valid results. For the same reason, the simplest model that will do the job should be used when working with PSpice.

Some circuits will have high-speed switching where the transition time is a function of the parameters of the transistor (and to a lesser degree the parameters of the circuit). Examples are astable multivibrators and Schmitt trigger circuits. In such circuits it is necessary to include in the transistor model those parameters which will accurately reflect the device switching time. Failure to do this can result in the switching time being so small that PSpice will be forced into using computing time steps that are extremely small, and as a result the transient analysis may not finish successfully.

11.3.1 Changing PSpice BJT Models

In order to obtain an accurate BJT model which gives realistic results in many circuits, the only parameter that needs to be changed is the ideal maximum forward beta, BF. It is the second parameter in the list of 55. The .MODEL line is used to set BF to other than its default value of 100. Typical values of BF range from 40 to 250. The element and model lines for three transistors with different forward betas would be

```
Q1  9 8 7 HIGH
Q2  6 5 4 MED
Q3  3 2 1 LOW
.MODEL HIGH  NPN(BF = 200)
.MODEL MED   NPN(BF = 120)
.MODEL LOW   NPN(BF = 40)
```

The switching time of a BJT is determined by its capacitances and its transit time. The capacitance parameters and their meanings are

CJE B-E zero-bias depletion capacitance
CJC B-C zero-bias depletion capacitance
CJS zero-bias collector-substrate capacitance

CJS is used only for IC transistors, and would be zero for a discrete transistor. CJC and CJE may be available from data sheets. Typical values are 1 or 2 pF.

The other parameter which affects switching time is the ideal forward transit time, TF. This is seldom available from data sheets, but there is a simple way to calculate a good approximation to this parameter:

$$TF = 1/(2*PI*fT)$$

where fT is called the common-emitter cutoff frequency, gain-bandwidth product or transition frequency. It is the frequency at which an extrapolation of the forward current gain on a graph of gain versus frequency falls to unity. Values of TF typically range from 0.1 ns to 20 ns for discrete transistors.

The default values of CJE, CJC and TF are zero. The element line and model line for a transistor with those parameters modified is

```
QOUTPUT  10  4  30   SPIFFY
.MODEL  SPIFFY  NPN(CJE = 1.2P  CJC = 0.8P  TF = 1.2N)
```

11.3.1.1 BJT Collector Characteristics

The following example uses the DC analysis function of PSpice to create the family of curves of collector volt-amp characteristics for a BJT, including using the VAF (forward Early voltage) model parameter.

Figure 11.2 shows a test circuit which is often used by students when they first encounter the bipolar junction transistor. A certain value of base current, IB, is set, and then the collector-emitter voltage, VCE, is varied over a range from 0 V to some maximum. For each value of VCE, the collector current, IC, is recorded. This produces data for a single plot of IC vs. VCE. Then IB is changed, and a new set of IC vs. VCE data is taken. When a graph is made of these data, a family of curves, called the collector volt-amp characteristics, results.

The PSpice input file, Fig. 11.3, uses the .MODEL line to modify the PSpice default transistor so that the forward beta, or DC current gain, is set to 50, and the forward Early voltage is set to -60 V. The effect of base-width modulation on the collector current, which alters the slope of the IC vs. VCE family of curves, is expressed by the Early voltage. In

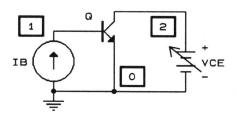

Figure 11.2 BJT Test Circuit.

```
BJT-FAM.CIR
*          produces a family
*          of collector volt-amp
*          characteristic curves
IB  0  1  DC  0
Q   2  1  0  TYPE1
.MODEL  TYPE1  NPN(BF = 50  VAF = 60)
VCE  2  0  DC  0
.DC VCE  0  20  .1  IB  0  1E-3  2E-4
.PROBE
.END
```

Figure 11.3 Input File BJT-FAM.CIR.

this example, the Early voltage is -60 V; however, in the .MODEL line it must be expressed as a positive value.

The two DC sources in the input file BJT-FAM.CIR are a current source (IB) and a voltage source (VCE). In the .DC control line, nested sweeps occur: IB is set to 0 uA, and VCE sweeps from 0 V to 20 V. Then IB is increased to 200 uA, and

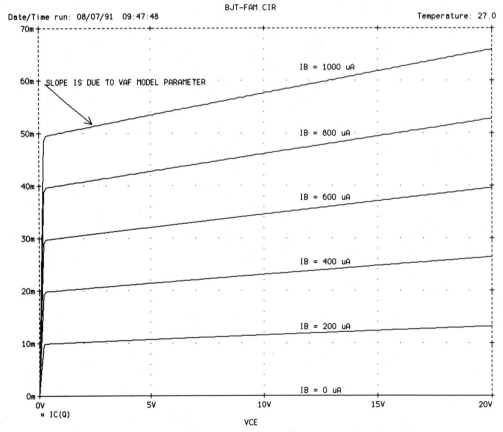

Figure 11.4 Family of BJT Collector Characteristics.

VCE again sweeps from 0 V to 20 V; this is repeated until IB reaches 1000 uA, producing six nested sweeps.

Figure 11.4 shows the family of collector characteristics for this transistor, with the progressively-increasing slope of each curve in the family clearly evident. The change in collector current between each 200 uA step of base current can be seen to be caused by the forward beta of 50. In Fig. 11.5 we can see the graphical interpretation of the VAF parameter: the forward Early voltage is the intersection of extensions of the IC curves with the negative VCE axis (-60 V in this case).

By taking collector characteristics for a BJT in the laboratory, one can determine the forward beta and forward Early voltage empirically and use these parameters to create a reasonable model for the transistor.

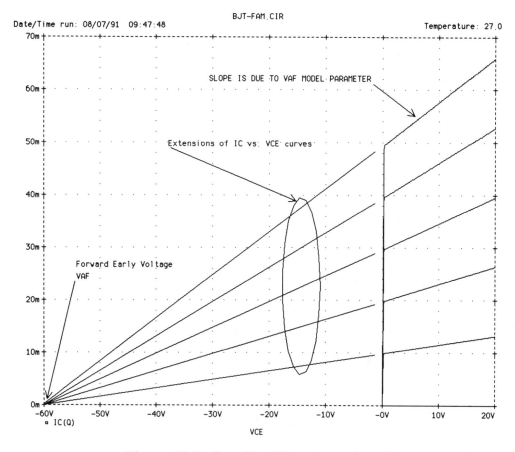

**Figure 11.5 Graphical Interpretation of
VAF (Forward Early Voltage) Parameter.**

11.3.2 Changing PSpice JFET Models

In order to obtain an accurate JFET model which gives realistic results in many circuits, the only parameter that needs to be changed is the transconductance parameter, BETA. It is the second parameter in the list of 21. The .MODEL line is used to set BETA to other than its default value of 1.0E-4. The BETA parameter has units of amp/(volt*volt); this is different from the transconductance, gm, that is normally available from manufacturers' data sheets. The unit for gm is amp/volt. If the channel length modulation parameter, LAMBDA, is neglected (its default value is zero) the relationship between gate-source voltage, VGS, and drain current, ID, in the PSpice JFET model is given by

$$ID = BETA*(VGS - VTO)^2$$

where VTO is the threshold voltage of the JFET. Since the transconductance, gm, is the derivative of drain current with respect to gate-source voltage, transconductance is

$$gm = \frac{d(ID)}{d(VGS)} = 2(BETA)(VGS - VTO)$$

Re-arranging the equation above for BETA gives

$$BETA = gm/(2(VGS - VTO)$$

Thus, if the data sheet for a JFET gives the parameters gm, VTO and VGS at a certain operating point, then the value of BETA can be calculated with the equation above. The element and model lines for two transistors with different transconductance parameters would be

```
J1  9 8 7  BIGGER
J2  6 5 4  SMALLER
.MODEL BIGGER    NJF(BETA = 5E-4)
.MODEL SMALLER PJF(BETA = 2E-4)
```

Notice that J1 is an N-channel JFET while J2 is a P-channel JFET.

11.3.3 Changing PSpice MOSFET Models

PSpice can simulate a MOSFET using four different models (Levels 1, 2, 3 and 4). The difference in the models is how the I-V characteristic is formulated. All four Levels share 28 model parameters. Three of the MOSFET models (Levels 1, 2 and 3) have an additional 24 parameters. The Level 4 MOSFET (Berkeley Short-Channel IGFET Model, or BSIM) has an additional 30 model parameters which are based on the fabrication process.

The fourth parameter for Levels 1, 2 and 3 is KP, the transconductance coefficient parameter. As with the JFET model, this parameter can easily be changed to any desired value other than its default value of 2E-5 by using the .MODEL line. Shown below are four lines which describe two MOSFETs with different transconductance coefficient parameters.

```
M1  9 8 7 6  BIGGER
M2  5 4 3 2  SMALLER
.MODEL BIGGER   PMOS(KP = 3E-5)
.MODEL SMALLER  NMOS(KP = 1E-5)
```

Note: For detailed information on parameters for the four Levels of MOSFET, including references, refer to the PSpice User's Guide, available from MicroSim Corporation.

11.4   USE OF Parts FOR DIODES

A major challenge facing a circuit designer when using a circuit simulator such as PSpice is to find models for the semiconductors in the circuit that are accurate. That is, the models must be good enough to allow the simulator to predict how the actual circuit, once constructed, will behave. If you have a model for a device, you <u>may</u> be all set; if the model isn't a good model for the conditions under which you are simulating the circuit behavior (e.g., large-signal operation with a small-signal model), a better device model is needed.

A PSpice option called Parts generates device models that are in a format immediately usable by PSpice. The production version of Parts will produce models for diodes, bipolar transistors, JFETs, power MOSFETs, bipolar and FET op-amps, and voltage comparators, while the evaluation version works only for diodes.

The only thing you must do before running Parts is to make sure there is a file in your PSpice subdirectory that will tell Parts what hardware (monitor and printer) you have connected to your PC. This file is called PSPICE.DEV (which will work for Probe and Parts), or PARTS.DEV (which will work for Parts if PSPICE.DEV doesn't exist). If you have a file

PROBE.DEV which works well with Probe, simply copy PROBE.DEV to PARTS.DEV. If not, refer to Chap. 7 which describes how to create PROBE.DEV, and make a file called PARTS.DEV in the same way.

Parts is menu-operated and very user-friendly, so let's get right into it. In your PSpice subdirectory, type PARTS and hit ENTER. Touch the "1" key, thus selecting diodes, then type in a name for your diode and hit ENTER, and the graph and menu shown in Fig. 11.6 should appear. This is the first screen in Parts, called Forward Current, and it has an I vs. V curve for the PSpice default diode and a menu at the bottom. The diode parameters IS, N, RS, IKF, XTI and EG are shown in a window. You can change any one of the parameters by selecting the "Model_parameters," or you can enter empirical voltage-current data by selecting "Device_curve." Doing either will produce a new display of the diode I vs. V curve with the altered parameters.

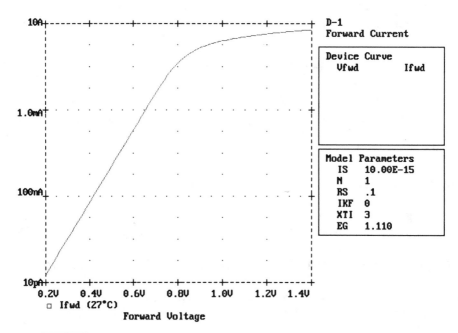

Exit **Next_set** Screen_info Device_curve Model_parameters Trace X_axis
Y_axis Hard_copy

Figure 11.6 Forward Current Screen for Parts.

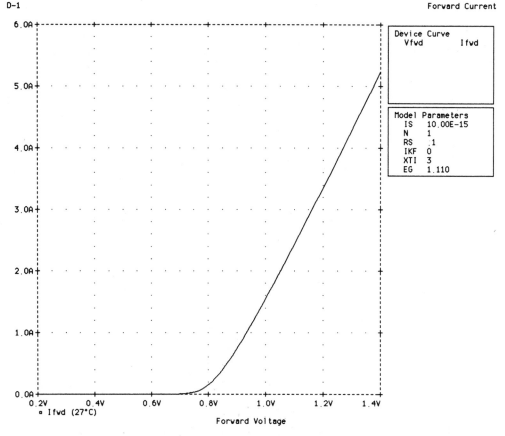

Forward Voltage

**Figure 11.7   Forward Current vs. Forward
Voltage Graph.**

If you enter empirical data, Parts will calculate the
PSpice parameters based on that data. Thus you can create
your own model of a diode you have measured in the laboratory.
Figure 11.7 is much the same as Fig. 11.6, except that the
y-axis has been made linear (vs. logarithmic). The "knee" of
the I vs. V graph is now evident. Notice that the temperature
is 27 degrees C. In Fig. 11.8 two additional curves have been
added by using "Trace": one for 0 degrees C, and one for 100
degrees C. In addition, "X__axis" was used to set the x-axis
limits.

Figure 11.9 was obtained by selecting "Next__set," which
brings up the Junction Capacitance screen. Now, the model pa-

Chap. 11   How to Change Semiconductor Models   **157**

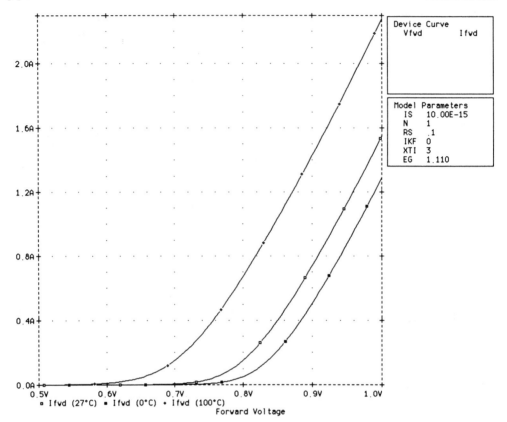

**Figure 11.8 Graph of Forward Current vs.
Forward Voltage at Three Temperatures.**

rameters shown are CJO, M, VJ and FC, all of which affect the
junction capacitance when the diode is reverse-biased.   Choosing "Next_set" again brings us to the Reverse Leakage screen,
Fig. 11.10, where the model parameters ISR and NR are displayed.   The next screen, Fig. 11.11, shows the Reverse Breakdown screen with model parameters BV and IBV. Here, the Device Data window allows you to enter Zener diode parameters obtained by laboratory measurement or from manufacturer's literature and have the PSpice diode model parameters calculated by
Parts.

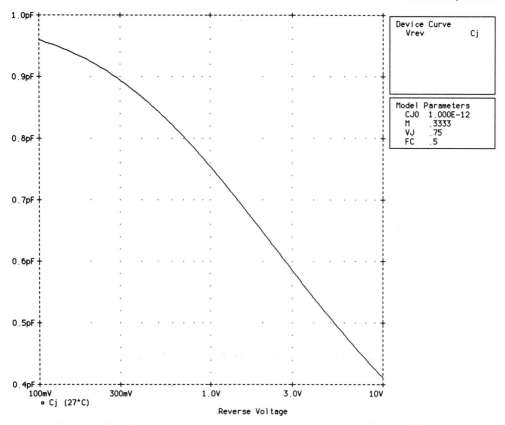

Figure 11.9 Junction Capacitance Screen.

The last screen is Reverse Recovery, Fig. 11.12, which is an indication of charge storage time in the junction, based on the diode model parameter TT. Having gotten to this point, you have had the opportunity to change 15 model parameters, and choosing "Next_set" gets you to the Exit screen. The Exit screen allows you to go through the different screens again, or to Exit Parts. The moment you exit, two files are created: a .MOD file which contains in ASCII text the PSpice .MODEL line for your diode, and a .MDT file which is used by Parts if

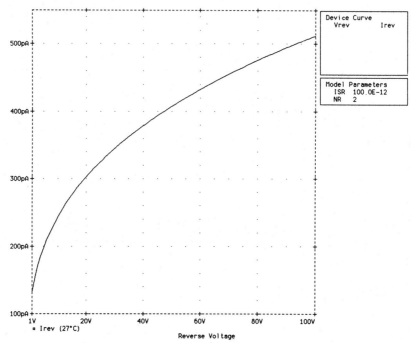

Figure 11.10 Reverse Leakage Screen.

Figure 11.11 Reverse Breakdown Screen.

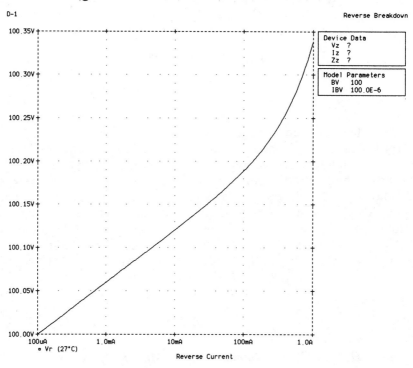

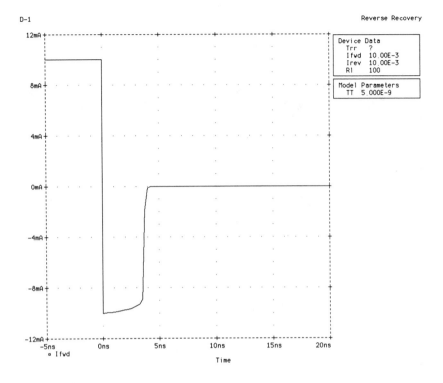

Figure 11.12 Reverse Recovery Screen.

you use Parts again with the same device name. Figure 11.13 is the file D-1.MOD, containing a comment line which tells the date, time and Parts version number, and two lines which have the 15 diode model parameters. This file can be made part of a PSpice input file when doing an analysis using this diode.

```
* D-1 model created using Parts version 4.05 on 08/07/91 at 16:17
*
.model D-1        D(Is=10f N=1 Rs=.1 Ikf=0 Xti=3 Eg=1.11 Cjo=1p M=.3333 Vj=.75
+                 Fc=.5 Isr=100p Nr=2 Bv=100 Ibv=100u Tt=5n)
```

Figure 11.13 Diode Model File D-1.MOD.

CHAPTER SUMMARY

PSpice recognizes five types of semiconductors (diode, BJT, JFET, MOSFET and GaAsFET) and has built-in models for each.

For many analyses, the built-in models need not be changed. If models need to be changed, generally only a few of the parameters have to be modified.

It is often difficult to obtain most semiconductor parameters for the PSpice model from manufacturers' data sheets or laboratory measurements. A PSpice option is Parts, which generates the .MODEL line parameters from the above information.

Chapter 12

SAMPLE CIRCUITS

12.1 WHY SAMPLE CIRCUITS ?

Perhaps the best reference to use when writing PSpice input files is previous files of a similar type. By looking back at successful PSpice analyses you can see how a particular type of analysis is specified, how to change temperature, how to specify a sub-circuit or how to print or plot a parameter of interest. The rest of this chapter is divided into five sections, each of which has examples of specific kinds of analysis. Considerable use is made of Probe for creating graphs of analysis results.

In some of the examples, you will encounter a parameter or option you may not have run into before. An excellent way to learn about parts of an input file is to try the analysis yourself, with and without the parameter or option. You will soon see its purpose for yourself.

163

12.3   DC analysis - one or more DC sources are varied and DC operating point is determined for each source value. Includes non-linear elements. Results may be printed and/or graphed.

12.4   AC analysis - a linear, small-signal analysis at one frequency, or over a range of user-specified frequencies. Results may be printed and/or graphed.

12.5   Transient analysis - specified output variables are determined as functions of time. Includes nonlinear elements. Results may be printed and/or graphed.

12.6   .TEMP - the circuit temperature may be set to any number of values (default temperature is 27 degrees C).

.FOUR - must be done in conjunction with a transient analysis. Performs a Fourier analysis of an output variable and gives amplitude and phase of first nine harmonics, as well as the DC component. The Fourier capability of Probe is illustrated as well.

.TF - does a small-signal transfer function analysis of a circuit from a specified input to a specified output. This results in the input resistance, output resistance and transfer function (voltage gain, current gain, transresistance or transconductance) being printed.

.OP - PSpice will solve for the DC operating point of a circuit including many transistor parameters, voltages and currents.

.STEP - a circuit element value is changed in steps; for each value of the circuit element, an analysis is performed.

.SENS - provides the small-signal DC sensitivities of one or more output variables, with respect to every parameter in the circuit.

.NOISE - must be done in conjunction with an AC analysis. PSpice will solve for the equivalent input and output noise of a circuit. Results may be printed and/or plotted.

12.2 NO ANALYSIS SPECIFIED

EXAMPLE 12.2.1, DCMESS.CIR

Figure 12.1 shows a DC circuit with one battery and 9 resistors. The problem is to find the DC voltage across RG, the 600 ohm resistor. The PSpice input file is shown in Fig. 12.2. It has element lines for each resistor and the battery. Even though no analysis is specified by a control line, PSpice will perform a small-signal analysis which will cause the voltage at each node and the voltage source current to be printed in the output file.

The short output file in Fig. 12.3 contains the small-signal bias solution, which gives the voltage at each node compared to ground. Since we need to know the voltage across resistor RG, it will be necessary to subtract manually the voltage at node 5 from the voltage at node 4, as follows

$$V_{RG} = V(4) - V(5) = 1.4474 - 0.8977 = 0.5497 \text{ V}$$

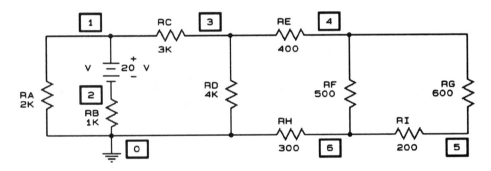

Figure 12.1 DC Circuit.

```
DCMESS.CIR    A CIRCUIT THAT COULD CAUSE A HEADACHE
V        1  2  20
RA       1  0  2K
RB       2  0  1K
RC       3  1  3K
RD       3  0  4K
RE       3  4  400
RF       4  6  500
RG       4  5  600
RH       6  0  300
RI       6  5  200
.OPTIONS NOPAGE
.END
```

Figure 12.2 Input File DCMESS.CIR.

```
**** 08/08/91 12:04:05 ********* Evaluation PSpice (January 1991) ************
    DCMESS.CIR   A CIRCUIT THAT COULD CAUSE A HEADACHE

    ****    CIRCUIT DESCRIPTION
*****************************************************************************

    V     1  2  20
    RA    1  0  2K
    RB    2  0  1K
    RC    3  1  3K
    RD    3  0  4K
    RE    3  4  400
    RF    4  6  500
    RG    4  5  600
    RH    6  0  300
    RI    6  5  200
    .OPTIONS NOPAGE
    .END

    ****    SMALL SIGNAL BIAS SOLUTION       TEMPERATURE =   27.000 DEG C

    NODE   VOLTAGE     NODE   VOLTAGE     NODE   VOLTAGE     NODE   VOLTAGE
    (   1)  11.3450  (   2)  -8.6545  (   3)   2.4001  (   4)   1.4474

    (   5)    .8977  (   6)    .7145


       VOLTAGE SOURCE CURRENTS
       NAME        CURRENT
       V         -8.655E-03

       TOTAL POWER DISSIPATION   1.73E-01  WATTS

          JOB CONCLUDED
          TOTAL JOB TIME            1.59
```

Figure 12.3 Output File DCMESS.OUT

EXAMPLE 12.2.2, VANDI.CIR

The DC circuit in Fig. 12.4 contains a voltage source, a current source and three resistors. The voltage across the 40 ohm resistor and the current through the 30 ohm resistor are to be found. Notice that a dead voltage source, V-AMP, is inserted in series with the 30 ohm resistor. It has no effect on the circuit operation and serves as an ammeter to give the current in that circuit branch.

Input file VANDI.CIR is shown in Fig. 12.5, and includes a .OPTIONS . . . NODE control line which causes an element node table to be printed. This table lists each node and the elements that are connected to those nodes. Using the NODE option can make it easier to find an error in a PSpice input

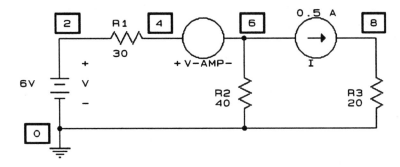

Figure 12.4 DC Circuit with "Ammeter".

file.   Figure 12.6 shows the output file VANDI.OUT, which
gives the node voltages.   The voltage across the 40 ohm resistor is simply V(6), or -5.1429 V, since the bottom of that resistor is at node 0.   The current through the 30 ohm resistor
is the current through V-AMP, which is 3.714E-01 A, or 371.4
mA.

Notice that the current through DC source V is listed as
-371.4 mA.   The reason for the negative sign is that PSpice
considers a current positive when it flows through a source
from the + node to the - node.   Since the current through DC
source V is from node 0 to node 2, it is negative (source V is
supplying power to the circuit).   The "Total Power Dissipation" listed, 2.23E+00 watts, is really just the power supplied to the circuit by the voltage source.   The current
source supplies 7.57 watts, so the power dissipated by all
three resistors in the circuit is 9.80 watts.

```
VANDI.CIR    VOLTAGE SOURCE AND CURRENT SOURCE DC CIRCUIT
*         the object is to solve for the voltage across
*         the 40-ohm resistor, and for the current through
*         the 30-ohm resistor.
V     2  0  6
R1    2  4  30
*         THE LINE BELOW IS A DEAD VOLTAGE SOURCE = AMMETER
V-AMP 4  6  0
R2    6  0  40
I     6  8  500M
R3    8  0  20
.OPTIONS NOPAGE NODE
.END
```

Figure 12.5 Input File VANDI.CIR.

```
**** 09/14/91 07:36:42 ********* Evaluation PSpice (January 1991) ************
  VANDI.CIR   VOLTAGE SOURCE AND CURRENT SOURCE DC CIRCUIT

  ****      CIRCUIT DESCRIPTION
********************************************************************************

  *       the object is to solve for the voltage across
  *       the 40-ohm resistor, and for the current through
  *       the 30-ohm resistor.
  V    2  0  6
  R1   2  4  30
  *       THE LINE BELOW IS A DEAD VOLTAGE SOURCE = AMMETER
  V-AMP 4  6  0
  R2   6  0  40
  I    6  8  500M
  R3   8  0  20
  .OPTIONS NOPAGE NODE
  .END

  ****      ELEMENT NODE TABLE

  0            V           R2          R3
  2            V           R1
  4            R1          V-AMP
  6            I           R2          V-AMP
  8            I           R3

  ****      SMALL SIGNAL BIAS SOLUTION      TEMPERATURE =   27.000 DEG C

  NODE   VOLTAGE     NODE   VOLTAGE     NODE   VOLTAGE     NODE   VOLTAGE
  (   2)   6.0000  (    4)  -5.1429  (    6)  -5.1429  (    8)  10.0000

       VOLTAGE SOURCE CURRENTS
       NAME          CURRENT
       V            -3.714E-01
       V-AMP         3.714E-01

       TOTAL POWER DISSIPATION    2.23E+00  WATTS

          JOB CONCLUDED
          TOTAL JOB TIME          1.54
```

Figure 12.6 Output File VANDI.OUT.

EXAMPLE 12.2.3, DC-CKT.CIR

The schematic diagram in Fig. 12.7 is a DC circuit with a linear voltage-controlled current source, called G. The current
through VCCS G is equal to five times VX, the voltage across
the 15 ohm resistor. The problem is to find the voltage across
the 6 ohm resistor and the current through the 25 ohm resistor. An ammeter (V-AMP, a dead voltage source) is put in series with the 25 ohm resistor for this purpose. In order to

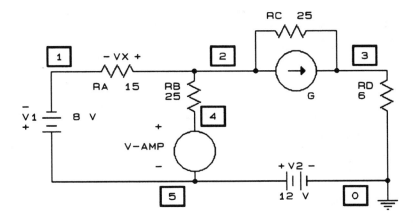

Figure 12.7 DC Circuit with Linear VCCS.

find the voltage across the 6 ohm resistor easily, ground (node 0) has been placed at its bottom.

The input file, Fig. 12.8, contains the control line .OP. This specifies an operating point analysis, which will cause the current through each source to be printed. The output file in Fig. 12.9 shows the voltage at node 3 to be 1.8907 V, and the current through V-AMP to be -318.2 mA. The negative polarity of the current through V-AMP tells us that current is flowing upward through the 25 ohm resistor, from node 4 to node 2.

```
DC-CKT.CIR     CONTAINS A VOLTAGE-CONTROLLED CURRENT SOURCE
V1      5  1  8
RA      1  2  15
RB      2  4  25
*          V-AMP is a dead voltagae source, used as an ammeter
V-AMP   4  5  0
*          G below is a voltage-controlled current source
G       2  3  2  1  5
RC      2  3  25
RD      3  0  6
V2      5  0  12
.OP
.OPTIONS NOPAGE
.END
```

Figure 12.8 Input File DC-CKT.CIR.

```
**** 08/08/91 13:17:49 ********* Evaluation PSpice (January 1991) ************
DC-CKT.CIR    CONTAINS A VOLTAGE-CONTROLLED CURRENT SOURCE

****     CIRCUIT DESCRIPTION
****************************************************************************

V1      5  1  8
RA      1  2  15
RB      2  4  25
*       V-AMP is a dead voltagae source, used as an ammeter
V-AMP   4  5  0
*       G below is a voltage-controlled current source
G       2  3  2  1  5
RC      2  3  25
RD      3  0  6
V2      5  0  12
.OP
.OPTIONS NOPAGE
.END

****     SMALL SIGNAL BIAS SOLUTION       TEMPERATURE =   27.000 DEG C

NODE   VOLTAGE    NODE   VOLTAGE    NODE   VOLTAGE    NODE   VOLTAGE
(   1)    4.0000  (   2)    4.0458  (   3)    1.8907  (   4)   12.0000
(   5)   12.0000

       VOLTAGE SOURCE CURRENTS
       NAME        CURRENT
       V1          -3.052E-03
       V-AMP       -3.182E-01
       V2          -3.151E-01
       TOTAL POWER DISSIPATION   3.81E+00  WATTS

****     OPERATING POINT INFORMATION      TEMPERATURE =   27.000 DEG C

**** VOLTAGE-CONTROLLED CURRENT SOURCES

NAME        G
I-SOURCE    2.289E-01

       JOB CONCLUDED

       TOTAL JOB TIME          1.65
```

Figure 12.9 Output File DC-CKT.OUT.

12.3   DC ANALYSIS

EXAMPLE 12.3.1, DI-VA.CIR

Figure 12.10 illustrates a test circuit which can generate a
volt-ampere curve for a diode. The diode is the PSpice default diode, so the .MODEL line contains only the model name
(PLAIN) and the type of semiconductor specification (D). All
the PSpice default values for diode parameters are therefore

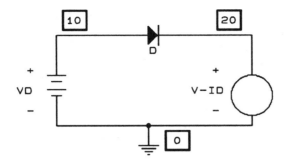

Figure 12.10 Diode Test Circuit.

used. The diode voltage is the same as the source (VD) voltage, and a dead voltage source (V-ID) has been put in series with the diode to measure current.

The input file is shown in Fig. 12.11. The diode voltage is stepped from 0.6 V to 0.81 V, in increments of 2 mV. The control line .PROBE will cause PSpice to produce an output file PROBE.DAT; a graph of diode current versus diode voltage can then be made by Probe.

Figure 12.12A shows the diode volt-amp characteristics. Using the cursor in Probe, at 600 mV the diode current is 118.8 uA, while at 810 mV the current has risen to 399.0 mA. If the Y-axis is made logarithmic rather than linear, a straight line graph results, due to the relationship between voltage and current in a diode (see Chap. 8). In Fig. 12.12B, Probe's ability to display a mathematical expression is used to graph the diode's static resistance, VD/ID, on the upper plot, and the diode's dynamic resistance d(VD)/d(ID) on the lower plot. It is interesting to note that at 640 mV, the static resistance is about 1.2K ohms, while the dynamic resistance is much lower, 46 ohms.

```
DI-VA        DIODE V-A CHARACTERISTIC WITH .DC ANALYSIS
VD      10   0  DC 0
D       10  20  PLAIN
V-ID    20   0   0
.DC VD  0.6  0.81  .002
.MODEL PLAIN D
.PROBE
.END
```

Figure 12.11 Input File DI-VA.CIR.

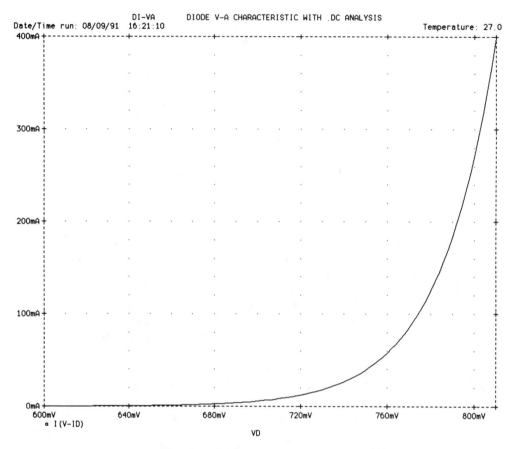

Figure 12.12A Diode Volt-Amp Characteristic.

EXAMPLE 12.3.2, FETINV.CIR

The junction field-effect transistor test circuit of Fig. 12.13 uses a PSpice default N-channel JFET. A graph of drain voltage versus gate-source voltage is desired. The gate-source voltage (VBATTERY) is stepped from -2.5 V to 0.5 V in steps of 20 mV.

Input file FETINV.CIR is shown in Fig. 12.14. The only specification for the JFET in the .MODEL line is NJF, so PSpice will use all the default parameters for JFETs. However, the element line for the JFET contains an area factor of 8. This means that the physical area of JFET J is 8 times larger than the default JFET, and several of the JFET parameters will be affected. See Section 8.4 for an explanation of

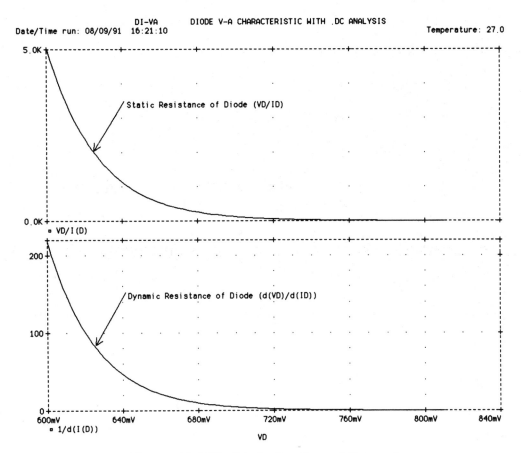

Static Resistance of Diode (VD/ID)

Dynamic Resistance of Diode (d(VD)/d(ID))

Figure 12.12B   Diode Static and Dynamic Resistances.

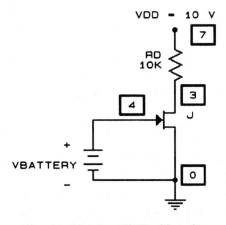

Figure 12.13   JFET Circuit.

```
FETINV.CIR    N-CHANNEL JFET INVERTER, TRANSFER CHARACTERISTIC
J        3  4  0  MOD1  8
.MODEL MOD1  NJF
VBATTERY 4  0  DC  0
VDD      7  0  10
RD       7  3  10K
.DC VBATTERY  -2.5  0.5  .02
.PROBE
.END
```

Figure 12.14 Input File FETINV.CIR.

the area factor. Notice that the element name of the JFET is
simply J. If there were another JFET in the circuit, it would
have to have another name, such as J2, JA, JBUFFER, etc.

Figure 12.15 is a Probe graph of drain voltage, V(3), versus gate-source voltage, VBATTERY. The drain voltage begins
to drop when VBATTERY reaches -2 V, and the JFET is saturated

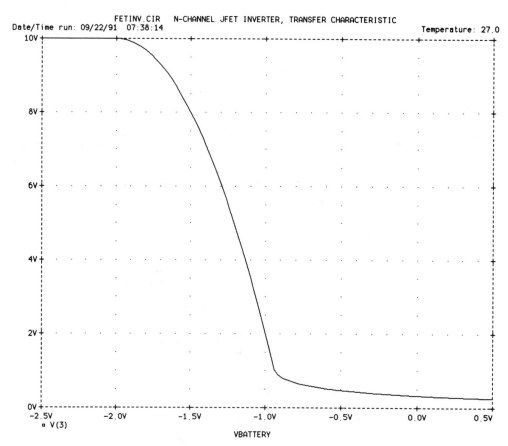

**Figure 12.15 JFET Inverter Transfer
Characteristic.**

when VBATTERY is above -1 V. The Probe cursor could be used to get exact numerical data, if needed.

EXAMPLE 12.3.3, TTLINV.CIR

Figure 12.16 is a schematic diagram of a TTL 7404 inverter. The output is connected to a 2K ohm pull-up resistor. In order to determine the transfer characteristic of this circuit, a DC source (VIN) will be stepped from 1.3 V to 1.7 V in increments of 1 mV. This is the range of input voltage within which the output changes from logical 1 to logical 0 at room temperature.

The input file in Fig. 12.17 shows that the diode uses the PSpice model, and the four transistors use the PSpice model except that forward beta is set to 50 (default value is 100). The .PROBE control line allows a graph of output voltage versus input voltage to be made using Probe.

The graph of Fig. 12.18 shows that the output voltage stays at nearly 5 V until the input voltage increases to about 1.45 V. As the input voltage rises from 1.45 V to 1.63 V, the output monotonically decreases, although with a discontinuity at about 1.53 V. For input voltages above 1.63 V, the output is less than 30 mV, indicating that transistor Q4 is well into saturation.

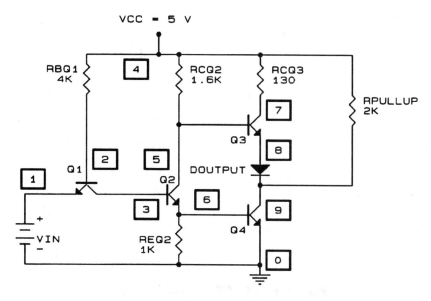

Figure 12.16 TTL Inverter Circuit.

```
TTLINV.CIR   7404 TTL INVERTER CIRCUIT, SWEPT DC INPUT
VCC      4  0  DC  5
VIN      1  0  DC  0
RBQ1     4  2  4K
RCQ2     4  5  1.6K
REQ2     6  0  1K
RCQ3     4  7  130
RPULLUP  4  9  2K
DOUTPUT  8  9  DI-MOD
Q1       3  2  1  Q-MOD
Q2       5  3  6  Q-MOD
Q3       7  5  8  Q-MOD
Q4       9  6  0  Q-MOD
.MODEL DI-MOD D
.MODEL Q-MOD  NPN(BF = 50)
.DC VIN  1.3  1.7  0.001
.PROBE
.END
```

Figure 12.17 Input File TTLINV.CIR.

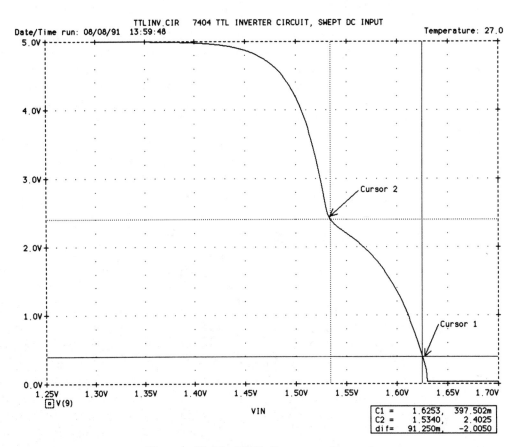

Figure 12.18 TTL Inverter Transfer Characteristic.

One note of caution: in a circuit which is bi-stable or has hysteresis, such as a Schmitt trigger (a TTL 7414 IC is one example), the use of DC analysis can be problematic. In the DC analysis, previous circuit values are not taken into consideration when calculating circuit conditions for the next input voltage. Since there are two possible stable output voltages for a given input voltage (depending on whether the input voltage is rising or falling), PSpice may fail to converge or may give incorrect results.

A better way to analyze a bi-stable circuit is to use a piece-wise linear (PWL) source with a triangular shape and perform a transient analysis. When a transient analysis is done PSpice does "remember" the previous circuit conditions when calculating at the next time increment.

12.4   AC ANALYSIS

EXAMPLE 12.4.1, LOPASS.CIR

The R-C circuit of Fig. 12.19 is connected to a 1 V sinusoid. It should have a half-power, or -3 dB, frequency of 500 Hz. PSpice AC analysis will be used to generate a Bode plot (graph of the magnitude and phase of the output voltage versus frequency) of the circuit response.

The input file LOPASS.CIR in Fig. 12.20 calls for the source voltage V to be stepped from 5 Hz to 0.5 MHz logarith-

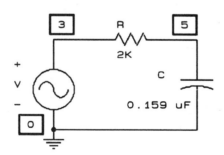

Figure 12.19   R-C Low-Pass Filter Circuit.

```
LOPASS.CIR    1-POLE R-C LOW PASS FILTER, AC ANALYSIS
V        3    0  AC  1
R        3    5  2K
C        5    0  0.159U
*        The RC values give a corner freq. of 500 Hz
.AC  DEC  20  5  0.5MEG
.PROBE
.END
```

Figure 12.20   Input File LOPASS.CIR.

mically, with 20 steps per decade of frequency. The output voltage (at node 5) will be plotted in decibels, using Probe.

Figure 12.21 contains a graph of VDB(5) indicating that the output voltage magnitude is essentially 0 dB at 5 Hz, falls to -3 dB at 500 Hz, and then linearly (on a logarithmic frequency axis) rolls off at 20 dB/decade. The Probe cursor shows that at 5 KHz, the output is -20.035 dB, and at 50 KHz (one decade of frequency above 5 KHz) the output is -39.992 dB.

The phase of the output voltage starts at 0 degrees, is -45 degrees at 500 Hz, and asymptotically approaches -90 degrees as frequency increases above 500 Hz.

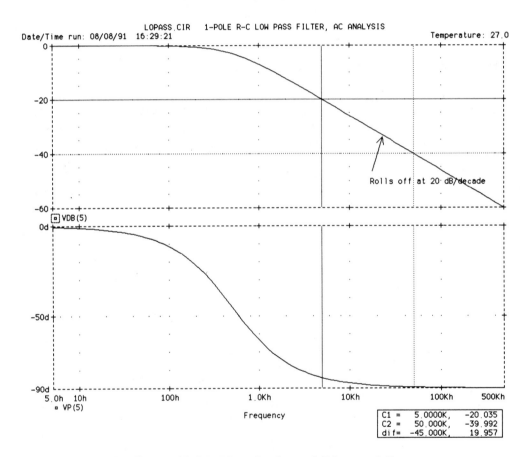

Figure 12.21 Magnitude and Phase of Low Pass Filter Response.

Sample Circuits   Chap. 12

EXAMPLE 12.4.2, WEIN.CIR

The Wein bridge network is frequently used to determine the frequency of a low frequency R-C oscillator. Figure 12.22 shows half of a Wein bridge, which is the interesting and more difficult to understand half (the other half is two resistors in series). In order to see the response of the network with the output at node 6, the input voltage will be swept logarithmically from 10 Hz to 1000 Hz with 50 frequencies per decade.

The input file in Fig. 12.23 sets the source voltage magnitude to 1 V. Notice that the two resistors are the same value, and the two capacitors have the same value, although they are specified differently.

As shown in Fig. 12.24, graphs of output voltage magnitude and phase, the phase becomes zero at 100 Hz, and is positive

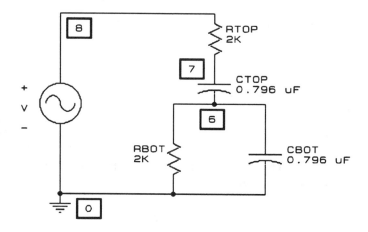

Figure 12.22 Wein Circuit.

```
WEIN.CIR     WEIN-BRIDGE NETWORK, EQUAL R, EQUAL C
V        8  0  AC  1
RTOP     8  7  2K
CTOP     7  6  796N
RBOT     6  0  2E3
CBOT     6  0  .796U
.AC  DEC  50  10  1000
.PROBE
.END
```

Figure 12.23 Input File WEIN.CIR.

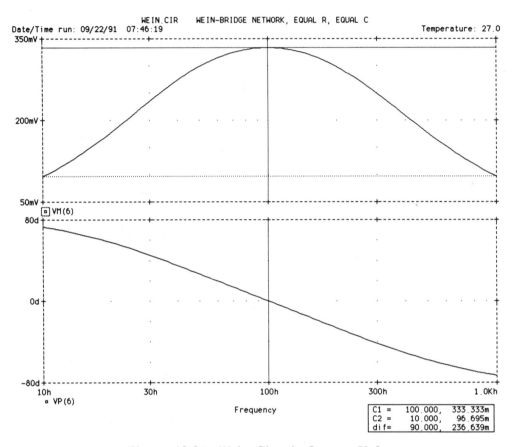

C1 =	100.000,	333.333m
C2 =	10.000,	96.695m
dif=	90.000,	236.639m

Figure 12.24 Wein Circuit Output Voltage and Phase.

below and negative above that frequency. The Wein circuit produces zero degrees of phase shift at only one frequency, which is also the frequency where the output amplitude is a maximum.

Since at 100 Hz the input voltage is 1 V, 0 degrees, and the output voltage is 0.3333 V at 0 degrees, the gain of the circuit is 1/3, 0 degrees. In order to make an oscillator using this circuit, a non-inverting gain (angle 0 degrees) of 3 (1/(1/3)) would have to be provided. This is true only if the resistors are equal and the capacitors are equal.

EXAMPLE 12.4.3, LCMATCH.CIR

The circuit in Fig. 12.25 would be troublesome at best to analyze by hand at a single frequency. It is a matching network,

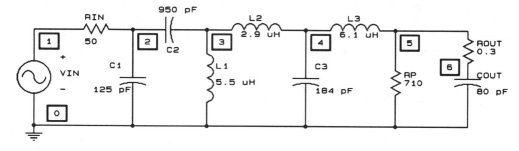

Figure 12.25 LCR Matching Network.

which is to be used at several frequencies. However, ROUT and COUT, the resistive and reactive parts of the load impedance, vary with frequency. For this reason the analysis will be at 2 MHz only.

The input file of Fig. 12.26 calls for an AC analysis at one frequency, with six voltages to be printed. The load resistance, ROUT, is between nodes 5 and 6, and the voltage across ROUT is to be printed as a magnitude and in decibel form.

If the 2 V input source (VIN and RIN) were connected to a matched resistive load of 50 ohms, the load would have 1 V across it (and 1 V would drop across RIN). Figure 12.27 is the output file which shows that the voltage across the load ROUT is 0.9861 mV, or -60.12 dB below 1 V. Obviously, this is not a high-efficiency matching network. Some of the voltages across reactances (VM(4) and VM(6), for example) exceed the

```
LCMATCH.CIR    L-C MATCHING NETWORK, SINGLE FREQ. AC ANALYSIS
VIN      1   0   AC  2
RIN      1   2   50
C1       2   0   125P
C2       2   3   950P
L1       3   0   5.5U
L2       3   4   2.9U
C3       4   0   184P
L3       4   5   6.1U
RP       5   0   710
ROUT     5   6   0.3
COUT     6   0   80P
*        Control Lines Follow
.AC  LIN  1  2MEG  2MEG
.PRINT AC  VM(2)  VM(2,3)  VM(4)  VM(6)
.PRINT AC  VM(5,6)  VDB(5,6)
.OPTIONS NOPAGE
.END
```

Figure 12.26 Input File LCMATCH.CIR.

Chap. 12   Sample Circuits   **181**

```
**** 08/08/91 17:03:30 ********* Evaluation PSpice (January 1991) ************
 LCMATCH.CIR    L-C MATCHING NETWORK, SINGLE FREQ. AC ANALYSIS

 ****       CIRCUIT DESCRIPTION
******************************************************************************

VIN     1  0  AC  2
RIN     1  2  50
C1      2  0  125P
C2      2  3  950P
L1      3  0  5.5U
L2      3  4  2.9U
C3      4  0  184P
L3      4  5  6.1U
RP      5  0  710
ROUT    5  6  0.3
COUT    6  0  80P
*       Control LInes Follow
.AC  LIN  1  2MEG  2MEG
.PRINT AC  VM(2)  VM(2,3)  VM(4)  VM(6)
.PRINT AC  VM(5,6)  VDB(5,6)
.OPTIONS NOPAGE
.END

 ****       SMALL SIGNAL BIAS SOLUTION          TEMPERATURE =   27.000 DEG C

 NODE   VOLTAGE     NODE   VOLTAGE     NODE   VOLTAGE     NODE   VOLTAGE
(    1)    0.0000  (    2)    0.0000  (    3)    0.0000  (    4)    0.0000
(    5)    0.0000  (    6)    0.0000

    VOLTAGE SOURCE CURRENTS
    NAME          CURRENT
    VIN           0.000E+00
    TOTAL POWER DISSIPATION   0.00E+00  WATTS

 ****       AC ANALYSIS                         TEMPERATURE =   27.000 DEG C

  FREQ         VM(2)        VM(2,3)      VM(4)        VM(6)
   2.000E+06   5.171E-01    2.516E+00    3.038E+00    3.270E+00

 ****       AC ANALYSIS                         TEMPERATURE =   27.000 DEG C

  FREQ         VM(5,6)      VDB(5,6)
   2.000E+06   9.861E-04   -6.012E+01

          JOB CONCLUDED
          TOTAL JOB TIME            1.87
```

Figure 12.27 Output File LCMATCH.OUT.

input voltage; this is indicative of some resonance effects occurring in the matching network.

One of the limitations of PSpice is shown here, which is that a load which is frequency-dependent cannot easily be described to PSpice. Antenna impedances are one common example of this type of load. Since the load impedance is a function of frequency, the input file must be edited (for load impedance) and a PSpice analysis must be done separately at each frequency.

EXAMPLE 12.4.4, SERIESAC.CIR

It is sometimes useful to be able to compare graphically the performance of two circuits. In this example the frequency response of two series resonant circuits which are identical except for the loss resistance in each will be compared. Probe will be used to plot the results on one frequency axis, to make the comparison easy.

Figure 12.28 illustrates the two circuits which are connected in parallel across a voltage source; in this way a single plot of two response curves can be made. A similar technique can be applied if a current source is the input by placing the two circuits in series.

Each circuit is resonant at 20 KHz; they differ in the series loss resistance, which determines quality factor Q and bandwidth. The source, V, will be swept linearly from 15 KHz to 25 KHz, and the voltage across the two load resistors will be plotted. The input file, Fig. 12.29, shows the control lines which will accomplish this.

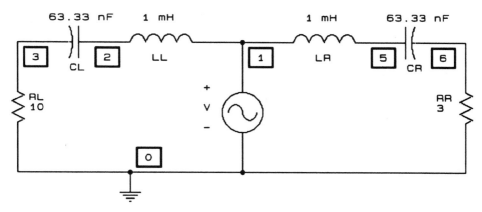

Figure 12.28 Two Series Resonant Circuits.

```
SERIESAC.CIR  TWO SERIES RESONANT CIRCUITS, ONE SOURCE
V        1  0  AC  1
*        LEFT SIDE SERIES CKT FOLLOWS, Q = 13
LL       1  2  1M
CL       2  3  63.33N
RL       3  0  10
*        RIGHT SIDE SERIES CKT FOLLOWS, Q = 42
LR       1  5  1E-3
CR       5  6  6.33E-8
RR       6  0  3
*        CONTROL LINES FOLLOW
.AC LIN 101 15K 25K
.PROBE
.END
```

Figure 12.29 Input File SERIESAC.CIR.

The Probe graph in Fig. 12.30 shows the graph of resistor voltage versus frequency for the two circuits. The substantial difference in the bandwidth of the two circuits is apparent, meaning that the left circuit, with a resistance of 10 ohms, has a lower Q and larger bandwidth than the right circuit which has a resistance of 3 ohms. The Probe cursors, on the V(3) trace, are located at the half-power frequencies (19.217 KHz and 20.813 KHz), with the bandwidth computed by Probe to be 1.5967 KHz.

EXAMPLE 12.4.5, DUBTUNE.CIR

The analysis of a circuit containing an RF transformer with a coefficient of coupling less than 1 is somewhat tedious, even at a single frequency. To get a frequency response plot is

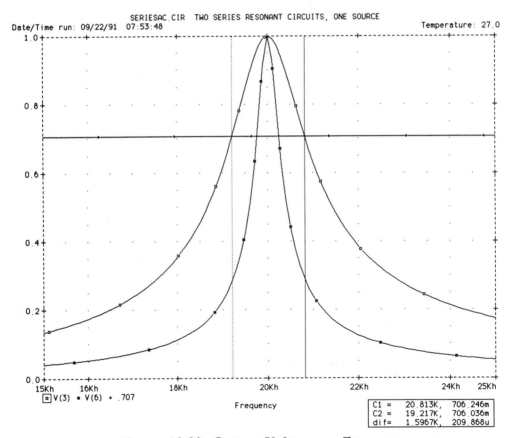

Figure 12.30 Output Voltage vs. Frequency for Two Resonant Circuits.

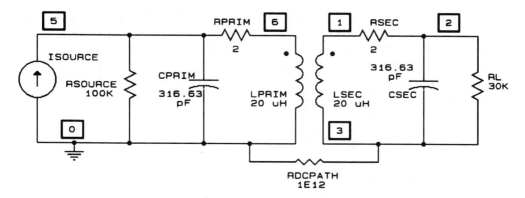

Figure 12.31 Double-Tuned RF Transformer.

not a task anyone would enjoy (or have time to do) by hand. Figure 12.31 illustrates such a circuit, which is fed by a current source.

One limitation of PSpice is apparent in this circuit: each node must have a DC path to ground so that a small-signal bias solution can be found. This is true for any circuit, even if the bias solution is meaningless since there are no DC sources in the circuit. Resistor RDCPATH is added to the circuit so that each node will have a DC path to ground. As shown in the input file, Fig. 12.32, RDCPATH has a value of 1E12 ohms, so its effect on the circuit performance is negligible.

Both primary and secondary windings of the transformer are resonant at 2 MHz, and the coefficient of coupling is greater than optimum coupling. This means that the transformer is over-coupled and will have a very undesirable frequency response.

```
DUBTUNE.CIR  DOUBLE-TUNED OVER-COUPLED RF TRANSFORMER
ISOURCE 0  5   AC  0.5M
RSOURCE 5  0   100K
CPRIM   5  0   316.63P
RPRIM   5  6   2
LPRIM   6  0   20U
KP-S       LPRIM  LSEC  0.10
LSEC    1  3   20U
RSEC    1  2   2
CSEC    2  3   316.63P
RL      2  3   30K
RDCPATH 3  0   1T
.AC LIN 201  1.8MEG  2.2MEG
.PROBE
.END
```

Figure 12.32 Input File DUBTUNE.CIR.

Figure 12.33, the output file, shows the classic double-peaked response, where the output has peaks at two frequencies, neither of which is the resonant frequency of 2 MHz. This is a good illustration of the concept of reflected impedance, in which the reactance of the secondary is reflected into the primary circuit and detunes the primary.

EXAMPLE 12.4.6, 2AC.CIR

A circuit with two AC sources is shown in Fig. 12.34. One is a voltage source, the other is a current source. The phasor voltage at node 5 is the unknown, over the frequency range of 1 KHz to 2 KHz.

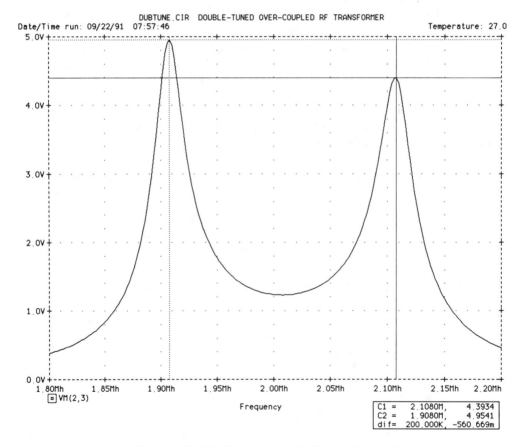

Figure 12.33 Response of Over-Coupled Double-Tuned Transformer.

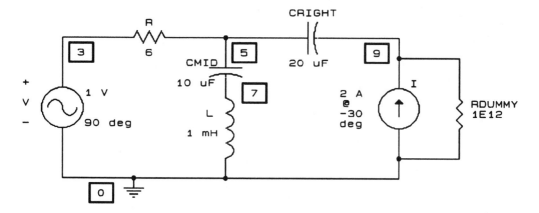

Figure 12.34 AC Circuit with Voltage and Current Source.

The circuit requires a dummy resistor (as did Example 12.4.5, above) to provide a DC path to ground for node 9. In the small signal bias solution, the capacitor and the AC current source are both open circuits. RDUMMY has a value of 1 teraohm (1E12 ohm), which should have little effect on circuit behavior. The .AC control line in Fig. 12.35 will cause an analysis to occur at 101 frequencies linearly spaced between 1 KHz and 2 KHz. Results of the AC analysis will be plotted by Probe.

Figure 12.36 is a graph of the phasor output voltage. The upper plot indicates that the voltage at node 5 is a minimum around 1.6 KHz, and that the phase changes abruptly at the same frequency. PSpice always expresses phase angles such that the magnitude of the phase angle is 180 or less. Thus, what appears to be an abrupt change in phase angle may simply be due to how it is expressed by PSpice.

```
2AC.CIR         L,C,R CIRCUIT WITH V AND I SOURCE
V       3  0  AC 1 90
I       0  9  AC 2 -30
R       3  5  6
L       7  0  1E-3
CMID    5  7  10U
CRIGHT  5  9  20U
RDUMMY  9  0  1T
*       Rdummy provides a DC path to ground for node 9
.AC LIN 101 1K 2K
.PROBE
.END
```

Figure 12.35 Input File 2AC.CIR.

Chap. 12   Sample Circuits   **187**

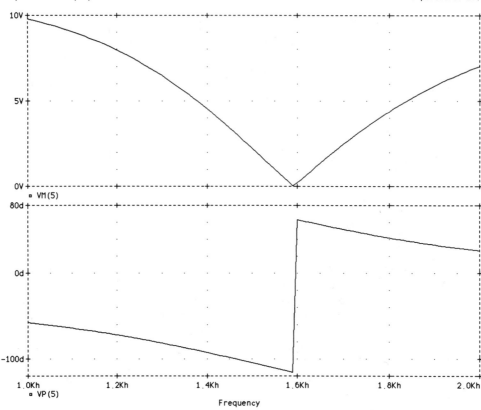

Figure 12.36 Magnitude and Phase of Output Voltage.

EXAMPLE 12.4.7, 2HP.CIR

Figure 12.37 shows two active filter circuits connected in parallel across an AC voltage source. Each circuit is a second-order high-pass filter; the first has Butterworth response, the second has Chebyshev response. The response is determined by the closed-loop gain of the operational amplifier.

A good AC model for an op-amp at low frequencies (when the op-amp is not operated near its slew-rate and gain-bandwidth product limits) is also shown in Fig. 12.37. The model includes an input resistance, which is rather large, and a voltage-controlled voltage source which provides the open-loop

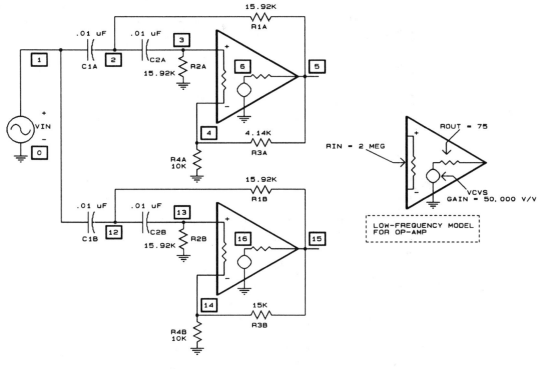

Figure 12.37 Two High-Pass Filters.

gain. The output voltage is the product of the differential input voltage and the open-loop gain.

The input file, Fig. 12.38, has comment lines to explain the function of the various element and control lines. This file would have been somewhat shorter if a subcircuit had been used to model the op-amp and the breakpoint frequency determining components. The frequency of the source is logarithmically swept from 100 Hz to 10 KHz, with 50 frequencies per decade.

The Probe graph of the filters' output, Fig. 12.39, indicates that the gain of the Chebyshev-response filter is higher than the Butterworth-response filter at all frequencies, with a noticeable peak at 1.122 KHz. While quantitative data could be obtained from this graph by using the cursor feature of Probe, the graph is of great use in qualitatively understanding the difference between these two filters. A .PRINT control line could be added to the input file to obtain quantitative information, if needed.

```
2HP.CIR   TWO ACTIVE HIGH-PASS FILTERS IN PARALLEL
VIN      1  0  AC  1
*              CIRCUIT "A" FOLLOWS
C1A      1  2  .01U
C2A      2  3  .01U
R1A      5  2  15.92K
R2A      3  0  15.92K
R3A      5  4  4.14K
R4A      4  0  10K
*              OP-AMP MODEL FOR CIRCUIT "A" FOLLOWS
RINA     3  4  2MEG
EA       6  0  3  4  50K
ROUTA    6  5  75
*              CIRCUIT "B" FOLLOWS
C1B      1  12  .01U
C2B      12  13  .01U
R1B      15  12  15.92K
R2B      13  0  15.92K
R3B      15  14  15K
R4B      14  0  10K
*              OP-AMP MODEL FOR CIRCUIT "B" FOLLOWS
RINB     13  14  2MEG
EB       16  0  13  14  50K
ROUTB    16  15  75
*              CONTROL LINES FOLLOW
.AC  DEC  50  100  10K
.PROBE
.END
```

Figure 12.38 Input File 2HP.CIR.

Figure 12.39 Bode Plot of Two High-Pass Filters.

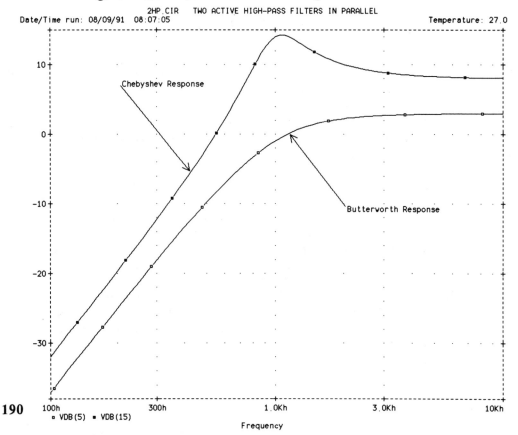

EXAMPLE 12.4.8, ACDIFAMP.CIR

A two-transistor BJT differential amplifier is shown in Fig. 12.40. It is fed by a voltage source at the base of Q1.

The input file, Fig. 12.41, includes two element lines for the transistors, but only one .MODEL line since the BJTs are matched and use the same model parameters. The model name is APPLE, which by itself means nothing until APPLE is defined in the .MODEL line. The model is of an NPN transistor, with a forward beta of 60 and significant junction capacitances representative of discrete devices, not transistors on an integrated circuit.

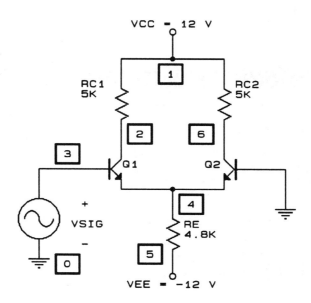

Figure 12.40 BJT Differential Amplifier.

```
ACDIFAMP.CIR     BJT DIFFERENTIAL AMPLIFIER, AC ANALYSIS
VSIG    3  0  AC  1
VCC     1  0  DC  12
VEE     5  0  -12
Q1      2  3  4  APPLE
Q2      6  0  4  APPLE
.MODEL APPLE NPN(BF=60  CJC=16P  CJE=30P)
RC1     1  2  5K
RC2     1  6  5K
RE      4  5  4.8K
.AC  DEC  5  100  1G
.PROBE
.END
```

Figure 12.41 Input File ACDIFAMP.CIR.

A Bode plot will be made by Probe, so the frequency is swept logarithmically from 100 Hz to 1 GHz and the output voltage (which is the gain, since the input voltage magnitude is one) will be plotted in decibels.

It is important ·to note that the AC analysis uses small-signal models for all devices. The input voltage of 1 V is hardly a small-signal input, and it causes a great deal more than a small-signal output. The output voltage in Fig. 12.42 shows an output voltage in the passband of 40.897 dBV, or 111 V. This is a ridiculous result and could not happen in the actual circuit because the power supplies would limit the output to about 12 Vpp. However, it is convenient to use 1 V as the input magnitude, since the output voltage is then the same ·as the circuit gain.

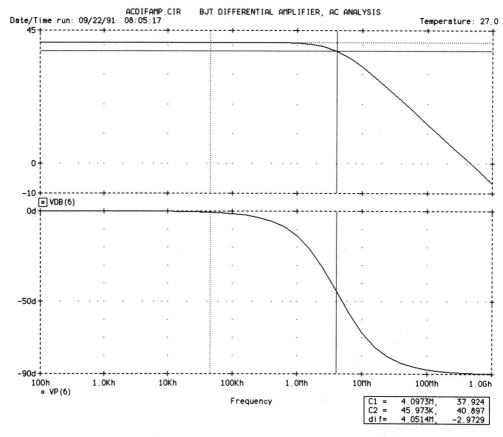

Figure 12.42 Bode Plot of Differential Amplfier Gain.

If the input were specified as 1 KV, the output would simply be 111 KV. Clearly, you must be careful when interpreting results of an AC analysis to be sure that the numbers are reasonable. If you wanted to see the effects of nonlinearities, such as clipping of the output due to the transistors becoming cut off and saturated, a transient analysis would have to be performed. The transient analysis does not use small-signal models as the AC analysis does.

The Bode plot shows the half-power frequency to be at 4.1 MHz, as indicated by the 3 dB (2.9729 dB is shown) drop in VDB(6) and the phase angle of -45 degrees near that frequency.

12.5 TRANSIENT ANALYSIS

EXAMPLE 12.5.1, TRIANG.CIR

A capacitor and resistor in series are connected to a triangle voltage source, as illustrated in Fig. 12.43. The triangle is made with the piece-wise linear (PWL) function, and varies between 0 V and 1 V, with a period of 2 ms.

Figure 12.44 lists input file TRIANG.CIR, which defines the triangle voltage from 0 ms to 6 ms only. Note that there

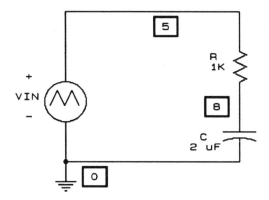

Figure 12.43 R-C Circuit, Triangle Input.

```
TRIANG.CIR    PIECEWISE LINEAR VOLTAGE INTO R-C LOWPASS FILTER
VIN     5  0  PWL(0 0  1M 1  2M 0  3M 1  4M 0  5M 1  6M 0)
R       5  8  1K
C       8  0  2U
.TRAN  .01M  6M
.PROBE
.END
```

Figure 12.44 Input File TRIANG.CIR.

are seven time-voltage pairs in the PWL definition. The transient analysis is from 0 ms to 6 ms, using time steps of 0.01 ms.

An examination of the graph of the capacitor voltage in the output file, Fig. 12.45, indicates that the capacitor voltage started at 0 V and had nearly reached steady-state in 6 ms. The average value of the triangle input voltage is 0.5 V.

For analyses that require many cycles of a periodic triangle waveform, using a PWL source can become rather tedious. A better way to generate a periodic triangle, pulse, sawtooth, etc. waveform is to use the PULSE function. When doing so, it is important to use a tiny (compared to the period) but non-zero value of pulse width. The rise time and fall time are set to equal half the period of the triangle waveform. A peri-

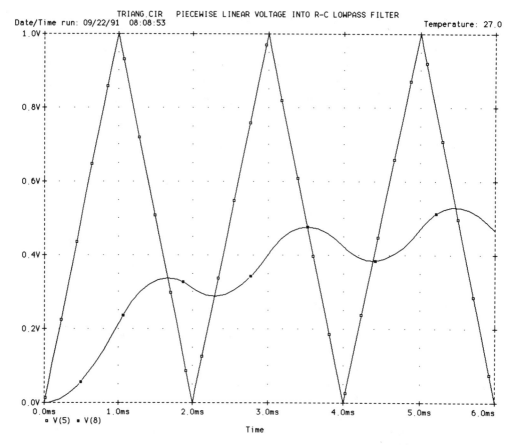

TRIANG.CIR   PIECEWISE LINEAR VOLTAGE INTO R-C LOWPASS FILTER
Date/Time run: 09/22/91 08:08:53   Temperature: 27.0

Figure 12.45 R-C Circuit Input and Output Voltages.

odic triangle waveform identical to the PWL triangle in this example could be produced by the following control line

VIN 5 0 PULSE(0 1 0 1M 1M 1N 2M)

Reading from left to right in the parentheses, the initial pulse value is 0 V, the pulsed value is 1 V, the delay time is 0, the rise time is 1 ms, the fall time is 1 ms, the pulse width is 1 ns (one millionth the rise or fall time) and the period is 2 ms. After 1000 cycles or so, the 1 ns pulse width, which should be 0, will start to introduce a slight error (a cumulative lengthening of the period by 0.0001% each cycle). However, the error is so slight that it doesn't much matter for any practical purpose.

EXAMPLE 12.5.2, LC.CIR

An inductor and capacitor in parallel (tank circuit) are connected to a pulse current source, as illustrated in Fig. 12.46. The tank circuit has a resonant frequency of 1007 Hz. This is a circuit that can be realized on paper only, as it is completely lossless. The pulse is very short, with a magnitude of 1 A.

The input file in Fig. 12.47 shows the PWL current source pulse to have a rise time, pulse width and fall time of 0.1 ms each. The pulse is delayed from time zero by 1 ms.

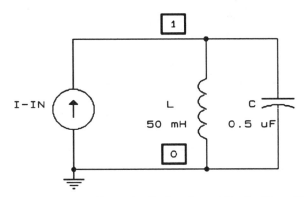

Figure 12.46 Current Pulse into Tank Circuit.

```
LC.CIR    PARALLEL L-C, NO LOSS, PULSE INPUT
I-IN    0  1  PWL(0  0  1M  0  1.1M  1  1.2M  1  1.3M  0)
L       1  0  50M
C       1  0  0.5U
.TRAN   .01M  4M  0  .005M
.PROBE
.END
```

Figure 12.47 Input File LC.CIR.

Examination of the Probe display of analysis results, Fig. 12.48, shows the current pulse on the upper plot and the voltage across the tank circuit on the lower plot. The pulse starts at 1 ms, and the tank voltage begins ringing at the same time. After the pulse ends at 1.3 ms, the tank voltage continues to ring with constant amplitude. This is due to the total absence of loss in the circuit; the tank will ring for as long as you care to do the transient analysis.

EXAMPLE 12.5.3, EKG-LP.CIR

A piece-wise linear source can be used to make any arbitrary waveform whatsoever, providing you have the patience to put all the information on the element line. This example con-

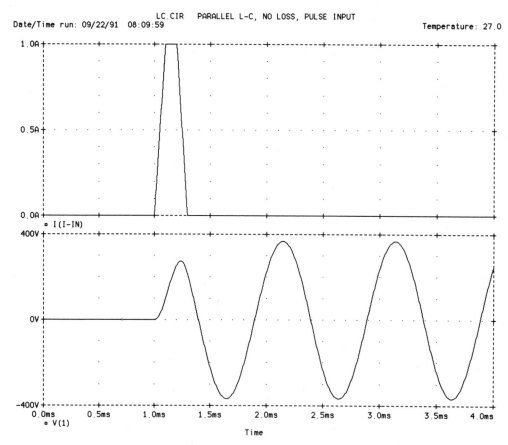

Figure 12.48 Input Current Pulse and Tank Circuit Voltage.

Sample Circuits   Chap. 12

tains an electrocardiogram (EKG) voltage over a time interval of 500 ms, with data points every 5 ms. The PWL independent source element "line" actually requires 16 lines, using the plus symbol in the first column to indicate a continuation of the previous line. The circuit, Fig. 12.49, is a single-pole R-C low-pass filter connected to the EKG voltage. It has a breakpoint frequency of 10 Hz.

The input file in Fig. 12.50 specifies that a transient analysis occur, with a time step of 1 ms. Figure 12.51 shows the analysis results. The effect of low-pass filtering on a complex waveform is very apparent, in that the filtered waveform is smoothed considerably and has a much smaller peak-to-peak value between 160 ms and 240 ms than the original EKG voltage.

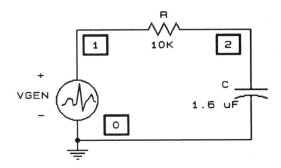

Figure 12.49 EKG into Low-Pass Filter.

```
EKG-LP.CIR  ELECTRO-CARDIOGRAM VOLTAGE INTO LOW-PASS FILTER
VGEN 1 0 PWL(0 -1 5M 2 10M 2 15M 2 20M 4 25M 5 30M 4
+ 35M 8 40M 8 45M 7 50M 8 55M 8 60M 6 65M 4 70M 3 75M -1
+ 80M -4 85M -4 90M -6 95M -9 100M -9 105M -7 110M -8
+ 115M -9 120M -7 125M -7 130M -8 135M -6 140M -7
+ 145M -9 150M -9 155M -8 160M -10 165M -9 170M -2
+ 175M 9 180M 25 185M 44 190M 54 195M 49 200M 34 205M 12
+ 210M -10 215M -25 220M -30 225M -30 230M -26 235M -20
+ 240M -14 245M -10 250M -5 255M -2 260M -3 265M -3
+ 270M -1 275M -1 280M -2 285M 0 290M 1 295M 0 300M 1
+ 305M 3 310M 2 315M 2 320M 4 325M 5 330M 3 335M 6
+ 340M 7 345M 7 350M 9 355M 11 360M 11 365M 12 370M 15
+ 375M 18 380M 18 385M 23 390M 25 395M 25 400M 28
+ 405M 32 410M 33 415M 35 420M 38 425M 38 430M 37
+ 435M 39 440M 36 445M 34 450M 31 455M 26 460M 22
+ 465M 19 470M 14 475M 10 480M 7 485M 5 490M 2
+ 495M 1 500M 1)
R      1  2 10K
C      2  0 1.6U
.TRAN 1M 500M 0 1M
.PROBE
.END
```

Figure 12.50 Input File EKG-LP.CIR.

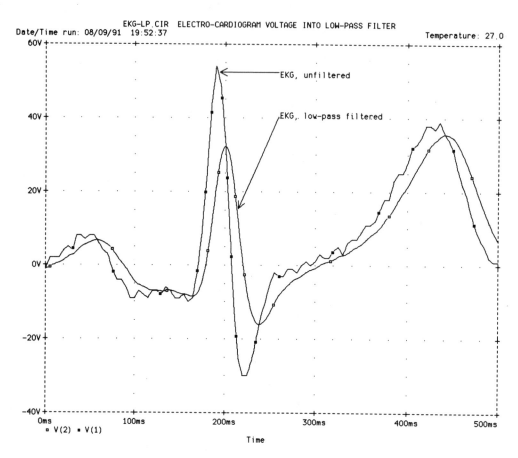

**Figure 12.51   Unfiltered and Low-Pass
Filtered EKG Voltage.**

EXAMPLE 12.5.4, PWRSUPLI.CIR

An interesting circuit to study with transient analysis is a
half-wave rectifier with a capacitor filter. The inrush current at the time of turn-on is not easily measured in the laboratory without some kind of storage device (analog or digital
storage oscilloscope). With PSpice transient analysis this
phenomenon can easily be examined in great detail, and insight
can be gained into power supply operation. Figure 12.52 is
the schematic diagram of the rectifier circuit, which includes
a dead voltage source in series with the diode to measure diode current.

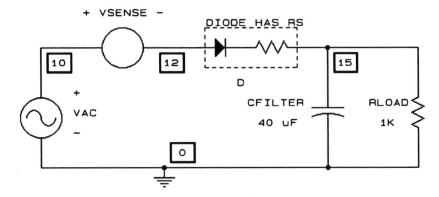

Figure 12.52 Half-Wave Rectifier Circuit.

The input file PWRSUPLI.CIR is shown in Fig. 12.53 Voltage source VAC is a 50 Hz sinusoid with an amplitude of 100 Vp. The voltage across the load resistor and the diode current will be plotted versus time by Probe.

Figure 12.54 plots the load voltage on the top and the diode current on the bottom. The load voltage rises from 0 V, peaks at nearly 100 V, and has the quasi-sawtooth shape characteristic of a poorly filtered power supply. The ripple voltage is about 33 Vpp. Using the cursor, we can see that the diode current during the first positive half cycle (inrush current) peaks at about 1.26 A, while during the second positive half cycle the maximum is 1 A. This is because the inrush current must charge the capacitor, which is initially at 0 V, and raise its voltage to 100 V. On subsequent positive half cycles the capacitor voltage never is less than 66 V, so less current is needed to raise its voltage back to 100 V.

```
PWRSUPLI.CIR   1/2 WAVE POWER SUPPLY
VAC 10 0 SIN(0 100   50)
*   VAC IS 100 VOLTS PEAK, 50 HZ
VSENSE 10  12  0
D   12  15  MODA
.MODEL  MODA  D(RS = 1)
CFILTER 15  0  40U
RLOAD   15  0  1K
.TRAN .1M 40M 0 .1M
.PROBE
.END
```

Figure 12.53 Input File PWRSUPLI.CIR.

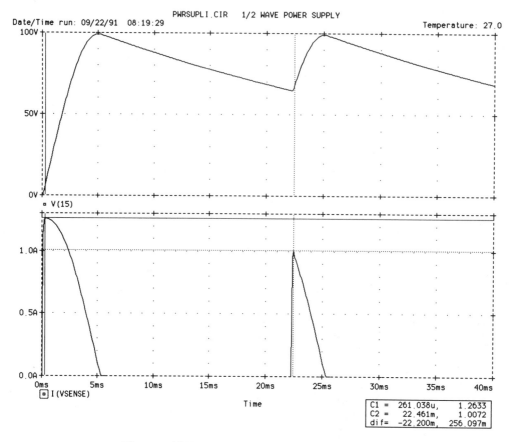

Date/Time run: 09/22/91   08:19:29   Temperature: 27.0

C1 =	261.038u,	1.2633
C2 =	22.461m,	1.0072
dif=	−22.200m,	256.097m

Figure 12.54 Power Supply Load Voltage and Diode Current.

EXAMPLE 12.5.5, CMOSNAND.CIR

A CMOS (complementary metal-oxide semiconductor) FET inverter NAND circuit is connected to two squarewave voltage sources in this example, and the output is determined as a function of time. The two squarewaves will provide all possible logic input conditions (11, 01, 10, 00 binary) to the NAND gate. The logic NAND gate is made with four enhancement MOSFETs; two are N-channel and two are P-channel.

Figure 12.55 shows the circuit, in which the NAND gate output is loaded resistively and capacitively.

The input file, Fig. 12.56, includes device capacitances and zero-bias threshold voltages in the .MODEL control lines.

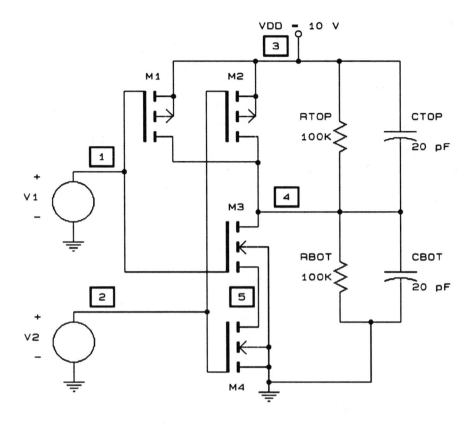

Figure 12.55 CMOS NAND Circuit.

```
CMOSNAND.CIR    4 TRANSISTOR NAND GATE, CMOS
M1  3  1  4  3  MOD1
M2  3  2  4  3  MOD1
.MODEL MOD1 PMOS(CGDO=10P  CGSO=10P  CBD=.05P  CBS=.05P)
M3  4  1  5  0  MOD2
M4  5  2  0  0  MOD2
.MODEL MOD2 NMOS(CGDO=10P  CGSO=10P  CBD=.05P  CBS=.05P)
VDD  3  0  10
RTOP  3  4  100K
CTOP  3  4  20P
RBOT  4  0  100K
CBOT  4  0  20P
V1  1  0  PULSE(0  10  1U  5N  5N  5U  10U)
V2  2  0  PULSE(0  10  1U  5N  5N  10U  20U)
.TRAN 0.05U  24U  0  .05U
.PROBE
.END
```

Figure 12.56 Input File CMOSNAND.CIR.

For enhancement-mode MOSFETs, VTO is positive for N-channel devices and VTO is negative for P-channel devices. The starting time of the pulses is delayed by 1 us, and the rise and fall times (5 ns) are quite small compared to the pulse widths (5 us and 10 us).

The analysis results are shown in Fig. 12.57. The upper plot contains the two input voltages and the lower plot is the NAND output voltage by itself. It can be seen that the NAND gate output is low only when both inputs are high, and that the rise and fall times of the output are significant. If desired, the cursor in Probe could be used to measure rise time and fall time of the output.

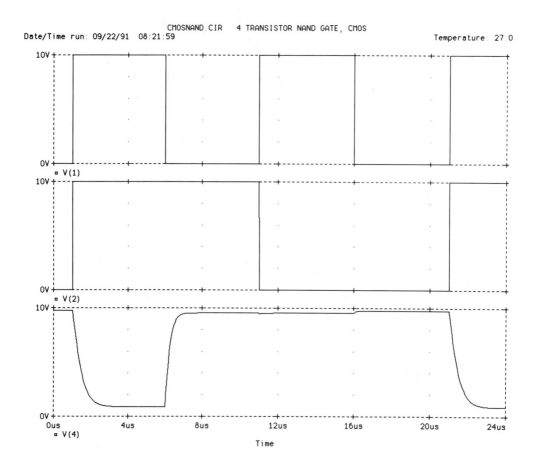

Figure 12.57 CMOS NAND Circuit Input and Output Voltages.

EXAMPLE 12.5.6, TTL.CIR

A 7404 TTL (transistor-transistor logic) inverter logic gate is shown in Fig. 12.58. It is fed by a TTL-compatible square-wave (well, nearly square). The output is connected to a pull-up resistor of 2K ohms.

The input file, Fig. 12.59, describes the input pulse as going from 0 V to 4V after a 10 ns delay, with a 2 ns rise

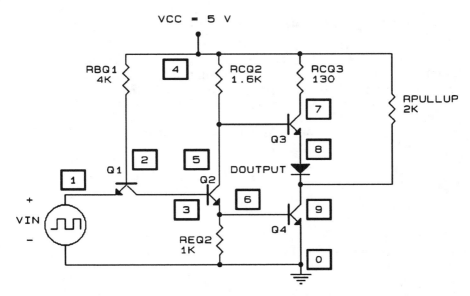

Figure 12.58 TTL Inverter.

```
TTL.CIR  TTL 7404 INVERTER GATE, PULSE INPUT
VCC 4 0 DC 5
VIN 1 0 PULSE(0 4 10N 2N 2N 20N 48N)
RBQ1 4 2 4K
RCQ2 4 5 1.6K
REQ2 6 0 1K
RCQ3 4 7 130
RPULLUP 4 9 2K
DOUTPUT 8 9 DMOD1
Q1 3 2 1 MOD1
Q2 5 3 6 MOD1
Q3 7 5 8 MOD1
Q4 9 6 0 MOD1
.MODEL MOD1 NPN(BF=50 BR=0.1 RB=70 RC=40 TF=.1N TR=10N
+  CJE=0.9P CJC=1.5P CCS=1P VA=50)
.MODEL DMOD1 D
.TRAN .2N  105N  0  .2N
.PROBE
.END
```

Figure 12.59 Input File TTL.CIR.

time and fall time, pulse width of 20 ns and a period of 48 ns. The transistor model MOD1 includes a substantial number of parameters which override the PSpice default parameters for a BJT.

The Probe display is a graph of input and output voltages (see Fig. 12.60). The output waveform is valid from a logic standpoint, and shows the distortions that are to be expected from a TTL logic gate operated at 20.8 MHz.

EXAMPLE 12.5.7, ASTABLE.CIR

Transient analyses on astable circuits and free-running oscillators can be a bit tricky to accomplish. One must pay careful attention to the initial conditions in the circuit; for

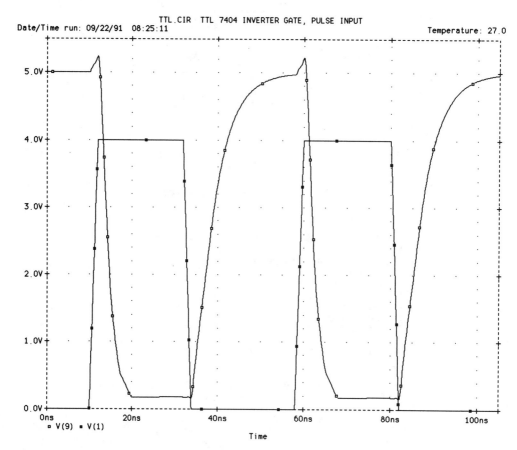

Figure 12.60 TTL Inverter Input and Output Voltages.

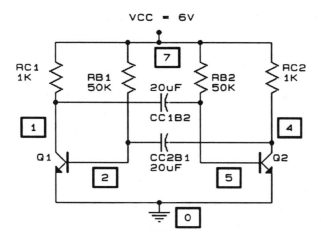

Figure 12.61 BJT Astable Multivibrator.

example, this circuit (see Fig. 12.61) has complete symmetry (at least on paper). In reality, at the time power is applied to the physical circuit one transistor will go to saturation and the other will become cut off. As far as PSpice is concerned, it doesn't know which transistor will win the race to conduction. In order to help PSpice with the transient solution, a good technique is to set some initial conditions which give PSpice a starting point for further analysis. This, of course, requires that you have an understanding of how the circuit starts and runs.

Such initial conditions are the five node voltages (expressed in the control line .IC . . .) specified in the input file in Fig. 12.62, where at time-zero Q1 is on (set by

```
ASTABLE.CIR  ASTABLE BJT CIRCUIT
*        OUTPUT IS A SQUAREWAVE
Q1  1  2  0 MOD1
Q2  4  5  0 MOD1
RC1  7 1  1K
RB1  7 2  50K
RC2  7 4  1K
RB2  7 5  50K
CC1B2  1  5 20U
CC2B1  4  2 20U
VCC 7 0 6
.MODEL MOD1 NPN (TF=10N CJC=1P CJE=1P)
.TRAN  0.01  2.5  0  .01  UIC
.IC V(1)=.1  V(4)=6  V(5)=-6  V(2)=0.8  V(7)=6
.OPTIONS  RELTOL=0.01
.PROBE
.END
```

Figure 12.62 Input File ASTABLE.CIR.

V(1)=.1 and V(2)=0.8) and Q2 is off (set by V(4)=6 and
V(5)=-6). Of course, in a physical circuit slight differences
in transistor parameters and/or stray circuit capacitances
might just as likely cause the opposite condition. Notice
that the .TRAN control line includes the UIC (use initial
conditions) specification. Also, the .OPTIONS control line
has RELTOL=0.01 included. This makes the relative tolerance
1% instead of the default value of 0.1%, which saves some computation time on a circuit such as this with very fast voltage
transitions.

The graph in Fig. 12.63 shows the collector voltage of Q1
in the lower plot and the base voltage of Q1 in the upper
plot, versus time. The collector voltage is essentially a
squarewave and the base voltage shows the exponential charging
of the 20 uF timing capacitor connected to it. Probe's cursor

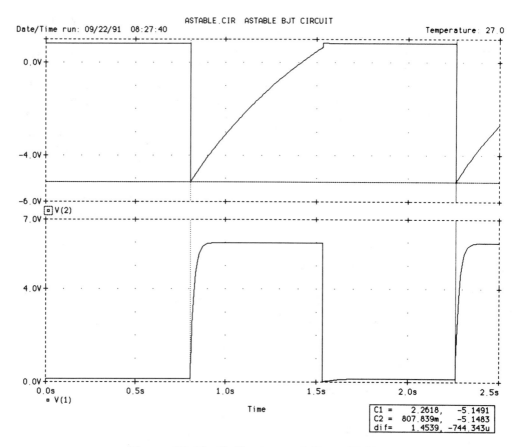

**Figure 12.63 Collector and Base Voltages
of Astable Multivibrator.**

used on the base voltage waveform allows the oscillation period to be measured as 1.4539 s. The behavior of the circuit once free-running oscillation has started confirms the validity of the original initial conditions.

EXAMPLE 12.5.8, ABSVAL.CIR

A real diode has an offset or barrier voltage (about 0.6 V for silicon) which must be overcome before forward conduction begins. This makes real diodes not usable by themselves to rectify low-level voltages such as audio signals. For rectifying small voltages another technique must be used to overcome this limitation. Figure 12.64 is a precision rectifier circuit, sometimes called an absolute value circuit. The very large open-loop gain of op-amp X1 is used to negate the diode offset voltage. The output voltage of the circuit is the absolute value of the input voltage; in other words, it is an ideal full-wave rectifier circuit.

The input file in Fig. 12.65 models the two operational amplifiers with subcircuits, in which the subcircuit contains a large input resistance, R-IN, and a big open-loop voltage gain, E-BIG. E-BIG is a VCVS (voltage-controlled voltage

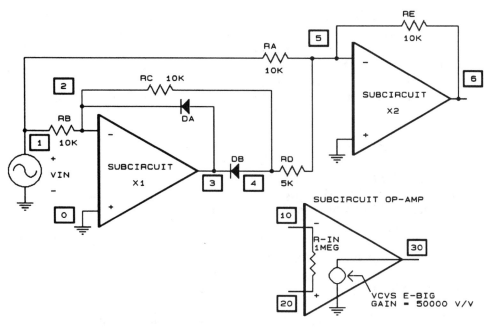

Figure 12.64 Precision Rectifier Circuit.

```
ABSVAL.CIR   ABSOLUTE VALUE (PRECISION RECTIFIER) CIRCUIT, FULL-WAVE
*            2 OP-AMPS IN CIRCUIT, SO SUB-CIRCUIT WILL BE USED FOR OP-AMP
.SUBCKT  OP-AMP  20  10  30
R-IN     20  10  1MEG
E-BIG    30  0  20  10  5E4
*            E-BIG IS A VCVS, GAIN OF 50,000 V/V
.ENDS OP-AMP
RA       1  5  10K
RB       1  2  10K
RC       2  4  10K
RD       4  5  5K
RE       5  6  10K
DA       3  2  ORDINARY
DB       4  3  ORDINARY
.MODEL  ORDINARY D
X1       0  2  3  OP-AMP
X2       0  5  6  OP-AMP
VIN      1  0  SIN(0  1  100)
.TRAN  .05M  20M  0  .05M
.PROBE
.END
```

Figure 12.65 Input File ABSVAL.CIR.

source) with a gain of 50,000 V/V which is controlled by the differential voltage across nodes 10 and 20 in the subcircuit.

The diodes are truly ordinary in that they are modeled by the PSpice default diode parameters, which makes them have an offset voltage typical of silicon. The input voltage is a 1 Vp, 100 Hz sinusoid, and the transient analysis occurs for 2 complete periods of the input.

In Fig. 12.66 the input voltage is plotted above and the output voltage below. The output voltage is indeed the absolute value of the input voltage, with no visible difference between input and output peak voltage caused by a diode. The accuracy of this circuit depends on both the large open-loop voltage gain of the op-amp and the exact matching of the resistors.

EXAMPLE 12.5.9, AM.CIR

Although PSpice has a built-in model for a frequency modulation generator (SFFM independent source), it does not have an amplitude modulation generator. By using a polynomial nonlinear voltage-controlled voltage source (VCVS), an AM generator can easily be created. The modulation index can changed, as can the carrier frequency and modulation frequency. While the SFFM source is limited to a pure sinusoid as the modulation, the AM generator could be modulated by any kind of waveform.

Figure 12.67 is the schematic diagram of the AM generator. V-CAR is the carrier sinusoid and V-MOD is the modulation volt-

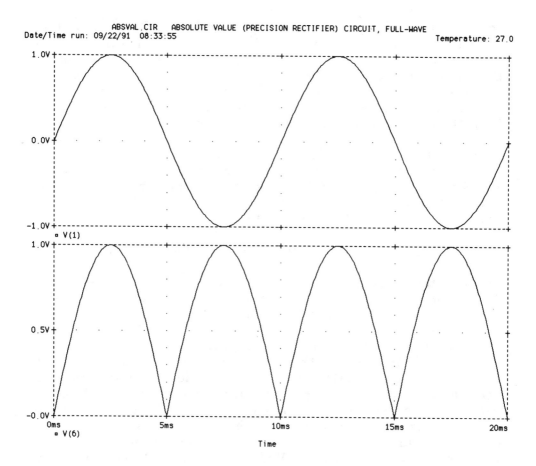

Figure 12.66 Precision Rectifier Input and Output Voltages.

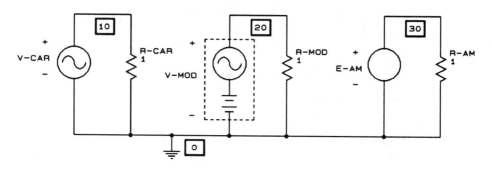

Figure 12.67 Amplitude Modulation Generator.

Chap. 12   Sample Circuits   **209**

age, which happens to be a sinusoid with a DC offset voltage. It could be a pulse, exponential, piece-wise linear or even SFFM source. Both of these generators will be used to control the output of the VCVS E-AM, across which the AM carrier will appear. Since PSpice requires that each node have more than one element connected to it, three dummy resistors (R-CAR, R-MOD and R-AM) are used to satisfy this requirement.

The input file in Fig. 12.68 specifies the carrier voltage to be a 10 Vp 40 KHz sinusoid and the modulation voltage to be 1 VDC in series with a 1 Vp, 2 KHz sinusoid. The polynomial VCVS E-AM multiplies V-CAR by V-MOD because the p4 coefficient (see Appendix C.1.2) is 1 and all other coefficients are 0. The equation for E-AM would then be

$$e(t) = (1 + 1 \sin(2*PI*2K*t)) \, (10 \sin(2*PI*40K*t)$$

which is the equation of a 100% modulated AM, or DSB-FC, carrier. The percent of modulation can be decreased by lowering the amplitude of the 2 KHz sinusoid from 1 Vp. The only change needed to change E-AM into a double sideband suppressed carrier is to change the DC offset voltage from 1 V to 0 V.

Figure 12.69A shows three plots: the 2 KHz modulating sinusoid on top, the 40 KHz carrier in the middle, and the AM carrier on the bottom. The amplitude modulation "envelope" is clearly visible on the AM carrier plot. Using the fast Fourier transform (FFT) feature of Probe on the AM carrier alone, a plot of amplitude vs. frequency is shown in Fig. 12.69B. The carrier amplitude of 10 Vp, and the lower and upper side frequencies of 5 Vp, are displayed. The lower and upper side frequencies are spaced 2 KHz below and above, respectively, the carrier frequency.

```
AM.CIR    AMPLITUDE MODULATION USING A POLYNOMIAL VCVS
*         Polynomial controlled sources are covered in Appendix C
V-CAR     10  0  SIN(0  10  40K)
*         V-CAR is 10 Vp 40 KHz sine, the carrier
R-CAR     10  0  1
V-MOD     20  0  SIN(1  1  2K)
*         V-MOD is 1VDC + 1 Vp 2 KHz sine, the modulation
R-MOD     20  0  1
E-AM      30  0  POLY(2)  10  0  20  0  0  0  0  0  1
R-AM      30  0  1
.TRAN  2U  4M  0  2U
.OPTIONS ITL5=0
.PROBE
.END
```

Figure 12.68 Input File AM.CIR.

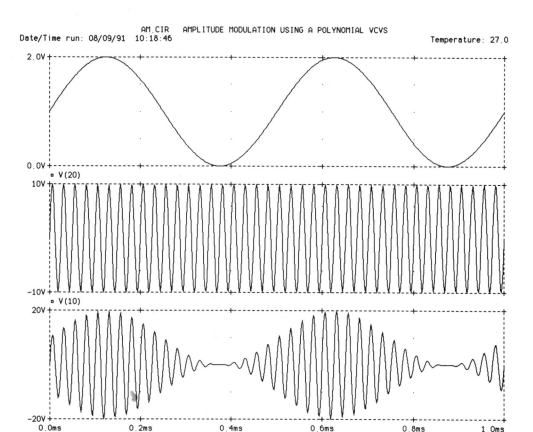

**Figure 12.69A Modulation, Carrier, and
Amplitude-Modulated Carrier Voltages.**

12.6 .TEMP, .FOUR, .TF, .OP, .STEP, .SENS & .NOISE CONTROL LINES

EXAMPLE 12.6.1, VARYTEMP.CIR

In order to examine the sensitivity of diode current to temperature, a diode with PSpice default parameters is connected
to a DC voltage source which can be stepped through a range using a DC analysis.   The DC analysis results can be used to
plot a volt-ampere characteristic of the diode.   Figure 12.70
shows the dead voltage source VSENSE connected in series to
measure diode current.

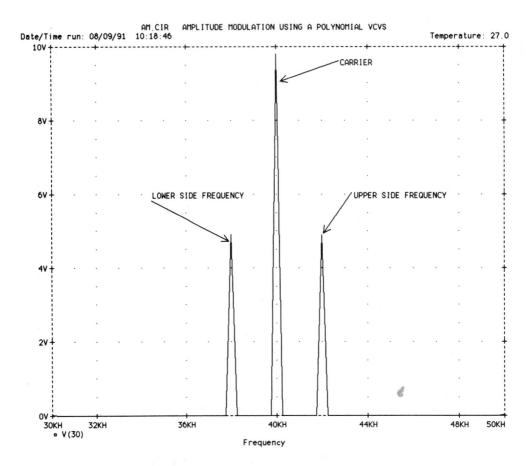

Figure 12.69B Fast Fourier Transform of Amplitude-Modulated Carrier Voltage.

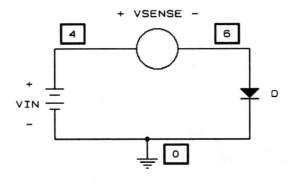

Figure 12.70 Diode Test Circuit.

```
VARYTEMP.CIR    PLOTS DIODE I-V CURVES AT DIFF TEMPS.
VIN  4  0  DC  0.5
D    6  0  TYPE1
VSENSE  4  6  0
.MODEL TYPE1 D
.DC VIN .65  .80  .001
.TEMP 0  50  100
.PROBE
.END
```

Figure 12.71 Input File VARYTEMP.CIR.

The input file, Fig. 12.71, shows that the DC source is stepped from 0.65 V to 0.80 V. The .TEMP control line specifies that the DC analysis will occur at a circuit temperature of 0 degrees C, and will be repeated at both 50 and 100 degrees C. For each analysis, a plot will be made of diode current versus diode voltage.

The traces in Fig. 12.72 may look unusual at first, because the current axis is logarithmic. The uppermost trace is

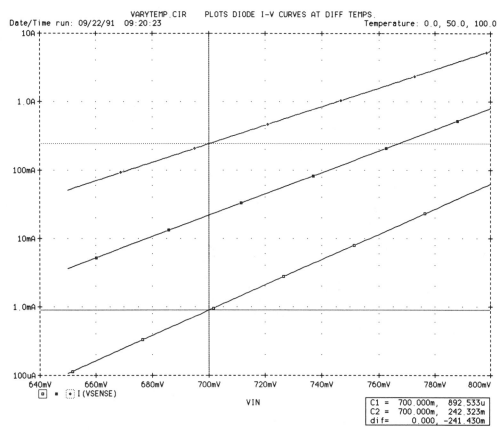

Figure 12.72 Diode Current vs. Voltage at Three Temperatures.

Chap. 12   Sample Circuits   **213**

the diode current at 100 degrees. Using the Probe cursors, we can determine that for a diode forward voltage of 0.7 V the diode current is 892.5 uA at 0 deg C and 242.3 mA at 100 deg C. The Parts software, described in Chap. 11, could also be used to make this kind of analysis.

EXAMPLE 12.6.2, DELTAT.CIR

The operation of a BJT differential amplifier with a constant current source, at an elevated temperature, is done using the .TEMP control line. Figure 12.73 shows the circuit, which includes a single-ended input voltage source VIN connected to the base of Q1.

The input file, Fig. 12.74, includes a .TF (transfer function) control line. This is explored in Ex. 12.6.4 as well. The transfer function will give the input resistance at VIN, the output resistance at node 3, and the differential-mode voltage gain from VIN to the collector of Q1 (node 3).

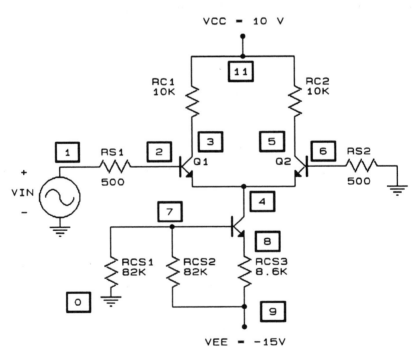

Figure 12.73 Differential Amplifier Circuit.

```
DELTAT.CIR   DIFF. AMP. OP. POINT AT 100 C
RS1  1  2  500
RS2  6  0  500
VIN  1  0  AC  0
RC1  11  3  10K
RC2  11  5  10K
VCC  11  0  10
VEE  9  0  -15
Q1  3  2  4  MOD1
Q2  5  6  4  MOD1
QCS  4  7  8  MOD1
RCS1  7  0  82K
RCS2  7  9  82K
RCS3  8  9  8.6K
.MODEL MOD1 NPN
.TF  V(3) VIN
.TEMP  100
.OPTIONS NOPAGE
.END
```

Figure 12.74 Input File DELTAT.CIR.

Output file DELTAT.OUT (Fig. 12.75) includes a listing of the BJT model parameters, a small-signal bias solution, and small-signal characteristics (the result of the .TF line) at the single temperature specified in the .TEMP line.   At 100 deg C the voltage gain is -54.70 V/V.

If a .OP control line were added to the input file, extensive operating point information about each transistor would also be included in the output file.   Adding other temperatures to the .TEMP control line would cause all analyses to be done at each specified temperature.

EXAMPLE 12.6.3, FOURIER.CIR

A Fourier analysis of a waveform can be done by PSpice in conjunction with a transient analysis.   The circuit of Fig. 12.76 is a half-wave rectifier with a resistive load.   If the correct fundamental frequency can be specified in the .FOUR control line, the amplitude and phase of the first nine harmonics present in the waveform will be calculated and printed along with the DC value of the waveform.   The fundamental frequency in this case is the input sinusoid frequency.

The input file, Fig. 12.77, shows the input source frequency to be 100 Hz, with a period of 10 ms and a peak voltage of 1 Vp.   The transient analysis is done for 100 ms, and the fundamental frequency is clearly 100 Hz.   The diode model is the PSpice default diode.

```
**** 08/09/91 12:36:19 ********* Evaluation PSpice (January 1991) ************
DELTAT.CIR   DIFF. AMP. OP. POINT AT 100 C

****       CIRCUIT DESCRIPTION
*****************************************************************************
RS1  1   2   500
RS2  6   0   500
VIN  1   0   AC  0
RC1  11  3   10K
RC2  11  5   10K
VCC  11  0   10
VEE  9   0   -15
Q1   3   2   4   MOD1
Q2   5   6   4   MOD1
QCS  4   7   8   MOD1
RCS1 7   0   82K
RCS2 7   9   82K
RCS3 8   9   8.6K
.MODEL MOD1 NPN
.TF  V(3)  VIN
.TEMP  100
.OPTIONS NOPAGE
.END

****       BJT MODEL PARAMETERS
                MOD1
                NPN
          IS  100.000000E-18
          BF  100
          NF  1
          BR  1
          NR  1

****       TEMPERATURE-ADJUSTED VALUES      TEMPERATURE = 100.000 DEG C
**** BJT MODEL PARAMETERS
NAME          BF         ISE        VJE        CJE        RE         RB
              BR         ISC        VJC        CJC        RC         RBM
              IS         ISS        VJS        CJS
MOD1      1.000E+02  0.000E+00  6.191E-01  0.000E+00  0.000E+00  0.000E+00
          1.000E+00  0.000E+00  6.191E-01  0.000E+00  0.000E+00  0.000E+00
          8.509E-13  0.000E+00  6.191E-01  0.000E+00

****       SMALL SIGNAL BIAS SOLUTION        TEMPERATURE = 100.000 DEG C
NODE    VOLTAGE      NODE    VOLTAGE      NODE    VOLTAGE      NODE    VOLTAGE
(   1)   0.0000   (   2)    -.0019   (   3)    6.2786   (   4)    -.6416
(   5)   6.2786   (   6)    -.0019   (   7)   -7.8082   (   8)   -8.4706
(   9) -15.0000   (  11)   10.0000

    VOLTAGE SOURCE CURRENTS
    NAME         CURRENT
    VIN        -3.721E-06
    VCC        -7.443E-04
    VEE         8.469E-04
    TOTAL POWER DISSIPATION   2.01E-02  WATTS

****       SMALL-SIGNAL CHARACTERISTICS
    V(3)/VIN = -5.470E+01
    INPUT RESISTANCE AT VIN =  1.828E+04
    OUTPUT RESISTANCE AT V(3) =  1.000E+04

        JOB CONCLUDED
        TOTAL JOB TIME          1.92
```

Figure 12.75 Output File DELTAT.OUT.

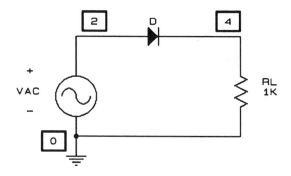

Figure 12.76 Half-wave Rectifier Circuit.

```
FOURIER.CIR      1/2-WAVE RECTIFIER W/FOURIER ANALYSIS
VAC  2  0  SIN(0  1.0  100)
D    2  4  SIMPLE
RL   4  0  1K
.MODEL   SIMPLE  D
.TRAN  .1M  100M  0  .1M
.FOUR  100  V(4,0)
.OPTIONS NOPAGE
.PROBE
.END
```

Figure 12.77 Input File FOURIER.CIR.

The PSpice output file, Fig. 12.78A, has a Fourier component table which lists the DC value to be 73.56 mV, and gives the amplitudes and phases of the first nine harmonics as well.

The graph in the Probe display, Fig. 12.78B, has two traces: the input sinusoid and the rectified output voltage. The x-axis was set to show the first 20 ms of time, so the shape of the output waveform could be clearly seen. At the peak of the output voltage the amplitude is 370 mV, indicating that the diode is dropping 630 mV. Note that the conduction angle of the diode is considerably less than 180 degrees. Figure 12.78C is the result of using Probe's Fourier transform in the X_axis menu, which graphs amplitude vs. frequency. The amplitudes of the DC term and the first five harmonics are shown; agreement between the data in Fig. 12.78A and this graph are excellent.

EXAMPLE 12.6.4, LH0005.CIR

A National Semiconductor LH0005 op-amp will be used to demonstrate the use of the .OP (operating point) and .TF (transfer

```
**** 08/10/91 08:11:12 ********* Evaluation PSpice (January 1991) ************
FOURIER.CIR     1/2-WAVE RECTIFIER W/FOURIER ANALYSIS
****      CIRCUIT DESCRIPTION

****************************************************************************

VAC 2 0 SIN(0 1.0 100)
D    2 4 SIMPLE
RL   4 0 1K
.MODEL SIMPLE D
.TRAN .1M 100M 0 .1M
.FOUR 100 V(4,0)
.OPTIONS NOPAGE
.PROBE
.END


****      Diode MODEL PARAMETERS
               SIMPLE
          IS   10.000000E-15


****      INITIAL TRANSIENT SOLUTION      TEMPERATURE =    27.000 DEG C
NODE   VOLTAGE      NODE   VOLTAGE      NODE   VOLTAGE      NODE   VOLTAGE
(   2)    0.0000  (    4)-862.4E-30

     VOLTAGE SOURCE CURRENTS
     NAME         CURRENT
     VAC          0.000E+00

     TOTAL POWER DISSIPATION   0.00E+00  WATTS


****      FOURIER ANALYSIS                TEMPERATURE =    27.000 DEG C
FOURIER COMPONENTS OF TRANSIENT RESPONSE V(4,0)

DC COMPONENT =    7.355792E-02


HARMONIC   FREQUENCY   FOURIER     NORMALIZED   PHASE       NORMALIZED
NO         (HZ)        COMPONENT   COMPONENT    (DEG)       PHASE (DEG)
  1        1.000E+02   1.340E-01   1.000E+00    1.869E-03   0.000E+00
  2        2.000E+02   9.988E-02   7.452E-01   -9.000E+01  -9.000E+01
  3        3.000E+02   5.728E-02   4.274E-01    1.800E+02   1.800E+02
  4        4.000E+02   2.018E-02   1.506E-01    8.998E+01   8.998E+01
  5        5.000E+02   2.473E-03   1.845E-02   -1.798E+02  -1.798E+02
  6        6.000E+02   9.620E-03   7.178E-02    9.001E+01   9.001E+01
  7        7.000E+02   6.563E-03   4.896E-02   -3.912E-02  -4.099E-02
  8        8.000E+02   6.429E-04   4.796E-03   -9.058E+01  -9.058E+01
  9        9.000E+02   2.926E-03   2.183E-02    5.272E-02   5.086E-02


     TOTAL HARMONIC DISTORTION =   8.768942E+01 PERCENT

          JOB CONCLUDED
          TOTAL JOB TIME          26.31
```

Figure 12.78A Output File FOURIER.OUT.

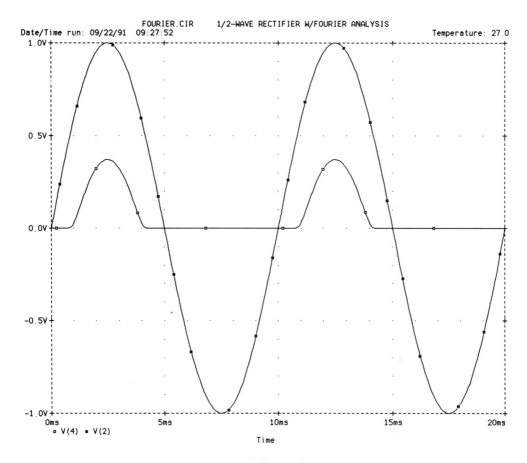

Figure 12.78B Rectifier Input and Output Voltages.

function) control lines.   The equivalent circuit of the LH0005 is shown in Fig. 12.79.   Not shown in the schematic are two in-put voltage sources, VIN1 and VIN2, which are connected to nodes 1 and 3, respectively.   Both VIN1 and VIN2 are dead sources, each with 0 V amplitude.

The input file, Fig. 12.80, contains the control line

.TF V(12) VIN1

which will give the DC small-signal input resistance at VIN1, the output resistance at node 12, and the open-loop DC voltage gain from VIN1 to the op-amp output at node 12.   The .OP con-

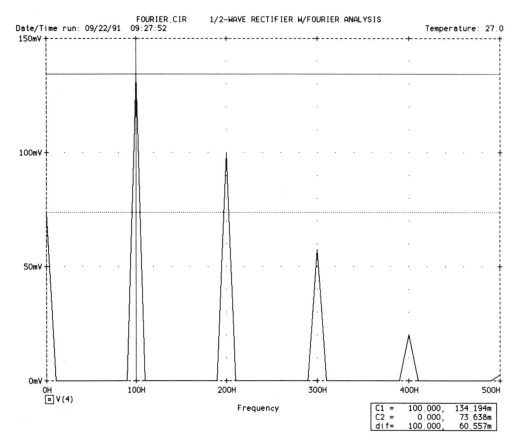

C1 =	100.000,	134.194m
C2 =	0.000,	73.638m
dif=	100.000,	60.557m

**Figure 12.78C Fourier Transform of
Rectifier Output Voltage.**

trol line causes PSpice to print detailed operating point information about semiconductor junction voltages, currents, and
other parameters. Since the transfer function input is VIN1,
the DC voltage gain that results will be the open-loop differential gain.

Figure 12.81 contains the output file which provides the
operating point information and transfer function results.
The open-loop differential gain is 3,754 V/V, which agrees
well with data sheet information. If you wanted to determine
the common-mode gain, the input file shown in Fig. 12.82 would
do that. The input voltage source VIN-CM is connected to node
1 and a 1E-6 ohm resistor, R-SHORT, connects node 1 with node

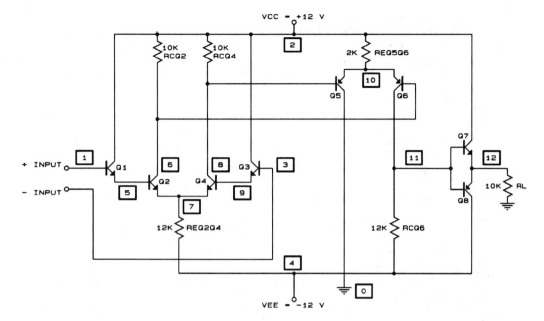

Figure 12.79 Equivalent Circuit of the LH0005 Op-Amp. (Courtesy of National Semiconductor Corp.)

```
LH0005.CIR    NATIONAL SEMICONDUCTOR OP-AMP IC
VCC     2  0  12
VEE     4  0  -12
Q1      2  1  5   MOD1
Q2      6  5  7   MOD1
Q3      2  3  9   MOD1
Q4      8  9  7   MOD1
RCQ2    2  6  10K
RCQ4    2  8  10K
VIN1    1  0  0
VIN2    3  0  0
REQ2Q4  7  4  12K
Q5      0  8  10  MOD2
Q6      11 6  10  MOD2
REQ5Q6  2  10 2K
RCQ6    11 4  12K
Q7      2  11 12  MOD1
Q8      4  11 12  MOD2
RL      12 0  10K
.MODEL  MOD1  NPN
.MODEL  MOD2  PNP
.OP
.TF  V(12)  VIN1
.OPTIONS NOPAGE
.END
```

Figure 12.80 Input File LH0005.CIR.

```
**** 08/09/91 13:06:11 ********* Evaluation PSpice (January 1991) ************
LH0005.CIR   NATIONAL SEMICONDUCTOR OP-AMP IC


****     CIRCUIT DESCRIPTION
*******************************************************************************

VCC      2  0  12
VEE      4  0  -12
Q1       2  1  5   MOD1
Q2       6  5  7   MOD1
Q3       2  3  9   MOD1
Q4       8  9  7   MOD1
RCQ2     2  6  10K
RCQ4     2  8  10K
VIN1     1  0  0
VIN2     3  0  0
REQ2Q4   7  4  12K
Q5       0  8  10  MOD2
Q6       11 6  10  MOD2
REQ5Q6   2  10 2K
RCQ6     11 4  12K
Q7       2  11 12  MOD1
Q8       4  11 12  MOD2
RL       12 0  10K
.MODEL   MOD1  NPN
.MODEL   MOD2  PNP
.OP
.TF   V(12)  VIN1
.OPTIONS NOPAGE
.END



****     BJT MODEL PARAMETERS
              MOD1              MOD2
              NPN               PNP
         IS  100.000000E-18  100.000000E-18
         BF  100               100
         NF  1                 1
         BR  1                 1
         NR  1                 1


****     SMALL SIGNAL BIAS SOLUTION       TEMPERATURE =   27.000 DEG C
   NODE   VOLTAGE     NODE   VOLTAGE     NODE   VOLTAGE     NODE   VOLTAGE
(    1)   0.0000  (     2)   12.0000  (     3)   0.0000  (     4)  -12.0000
(    5)   -.6335  (     6)   7.7086   (     7)  -1.3863  (     8)   7.7086
(    9)   -.6335  (    10)   8.4792   (    11)  -1.5325  (    12)   -.8231


    VOLTAGE SOURCE CURRENTS
    NAME        CURRENT
    VCC         -2.627E-03
    VEE          1.838E-03
    VIN1        -4.334E-08
    VIN2        -4.334E-08
    TOTAL POWER DISSIPATION    5.36E-02   WATTS
```

Figure 12.81 Output File LH0005.OUT.

```
****      OPERATING POINT INFORMATION      TEMPERATURE =   27.000 DEG C
**** BIPOLAR JUNCTION TRANSISTORS
```

NAME	Q1	Q2	Q3	Q4	Q5
MODEL	MOD1	MOD1	MOD1	MOD1	MOD2
IB	4.33E-08	4.38E-06	4.33E-08	4.38E-06	-8.71E-06
IC	4.34E-06	4.38E-04	4.34E-06	4.38E-04	-8.71E-04
VBE	6.33E-01	7.53E-01	6.33E-01	7.53E-01	-7.71E-01
VBC	-1.20E+01	-8.34E+00	-1.20E+01	-8.34E+00	7.71E+00
VCE	1.26E+01	9.09E+00	1.26E+01	9.09E+00	-8.48E+00
BETADC	1.00E+02	1.00E+02	1.00E+02	1.00E+02	1.00E+02
GM	1.68E-04	1.69E-02	1.68E-04	1.69E-02	3.37E-02
RPI	5.97E+05	5.91E+03	5.97E+05	5.91E+03	2.97E+03
RX	0.00E+00	0.00E+00	0.00E+00	0.00E+00	0.00E+00
RO	1.00E+12	1.00E+12	1.00E+12	1.00E+12	1.00E+12
CBE	0.00E+00	0.00E+00	0.00E+00	0.00E+00	0.00E+00
CBC	0.00E+00	0.00E+00	0.00E+00	0.00E+00	0.00E+00
CBX	0.00E+00	0.00E+00	0.00E+00	0.00E+00	0.00E+00
CJS	0.00E+00	0.00E+00	0.00E+00	0.00E+00	0.00E+00
BETAAC	1.00E+02	1.00E+02	1.00E+02	1.00E+02	1.00E+02
FT	2.67E+15	2.69E+17	2.67E+15	2.69E+17	5.36E+17

NAME	Q6	Q7	Q8
MODEL	MOD2	MOD1	MOD2
IB	-8.71E-06	-1.35E-11	-8.15E-07
IC	-8.71E-04	2.64E-11	-8.15E-05
VBE	-7.71E-01	-7.09E-01	-7.09E-01
VBC	9.24E+00	-1.35E+01	1.05E+01
VCE	-1.00E+01	1.28E+01	-1.12E+01
BETADC	1.00E+02	-1.95E+00	1.00E+02
GM	3.37E-02	0.00E+00	3.15E-03
RPI	2.97E+03	1.00E+14	3.17E+04
RX	0.00E+00	0.00E+00	0.00E+00
RO	1.00E+12	1.00E+12	1.00E+12
CBE	0.00E+00	0.00E+00	0.00E+00
CBC	0.00E+00	0.00E+00	0.00E+00
CBX	0.00E+00	0.00E+00	0.00E+00
CJS	0.00E+00	0.00E+00	0.00E+00
BETAAC	1.00E+02	0.00E+00	1.00E+02
FT	5.36E+17	0.00E+00	5.01E+16

```
****      SMALL-SIGNAL CHARACTERISTICS
      V(12)/VIN1 =  3.754E+03
      INPUT RESISTANCE AT VIN1 =  2.375E+06
      OUTPUT RESISTANCE AT V(12) =  4.151E+02

      JOB CONCLUDED
      TOTAL JOB TIME            2.30
```

Figure 12.81 (Continued).

3. This makes the input voltage be a common-mode input, so the transfer function gain that results is a common-mode gain. Its value is 1.132 V/V, which results in a common-mode rejection ratio of 3753/1.132, or 70.4 dB.

```
LH0005CM.CIR    NATIONAL SEMICONDUCTOR OP-AMP IC
*          COMMON-MODE INPUT, VIN-CM
VCC      2   0   12
VEE      4   0   -12
Q1       2   1   5   MOD1
Q2       6   5   7   MOD1
Q3       2   3   9   MOD1
Q4       8   9   7   MOD1
RCQ2     2   6   10K
RCQ4     2   8   10K
VIN-CM   1   0   0
R-SHORT  1   3   1U
*          R-SHORT IS A JUMPER (1E-6 OHM) TO TIE INPUTS TOGETHER
REQ2Q4   7   4   12K
Q5       0   8   10  MOD2
Q6       11  6   10  MOD2
REQ5Q6   2   10  2K
RCQ6     11  4   12K
Q7       2   11  12  MOD1
Q8       4   11  12  MOD2
RL       12  0   10K
.MODEL   MOD1 NPN
.MODEL   MOD2 PNP
.OP
.TF   V(12)  VIN-CM
.OPTIONS NOPAGE
.END
```

Figure 12.82   Input File LH0005CM.CIR.

EXAMPLE 12.6.5, DELTA-F.CIR

PSpice has the ability to change the value of a component in a
circuit, and will repeat one or more analyses for each value.
Figure 12.83 is a series resonant circuit, in which the capacitor is variable.   The resonant frequency of the circuit for
the values shown is 100.7 KHz.   In the input file, Fig. 12.84,
the capacitor is given a model name (CVARIABLE) which is defined in the .MODEL line as a capacitor (CAP).   The control
line beginning with .STEP causes the capacitor to be set to
0.8 of its nominal value, and the first AC analysis occurs.
Then the capacitor value is incremented by 0.1 of its nominal
value, to 0.9(1 nF), and the AC analysis is repeated.   In all,

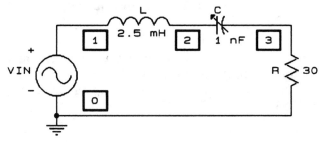

Figure 12.83   Resonant Circuit.

```
DELTA-F.CIR
*       series resonance with
*       varying capacitance
VIN 1  0  AC  1
L    1  2  .0025
C    2  3  CVARIABLE  1E-9
.MODEL  CVARIABLE  CAP
.STEP  CAP  CVARIABLE(C)  .8  1.2  .1
R    3  0  30
.AC LIN  281  85K  120K
.PROBE
.END
```

Figure 12.84 Input File DELTA-F.CIR.

five AC analyses are done as C is changed from 0.8 nF to 1.2 nF.

In Fig. 12.85 the five curves of V(3) vs. frequency are presented, showing the effect of changing the capacitor value on resonant frequency. It is apparent from the display that

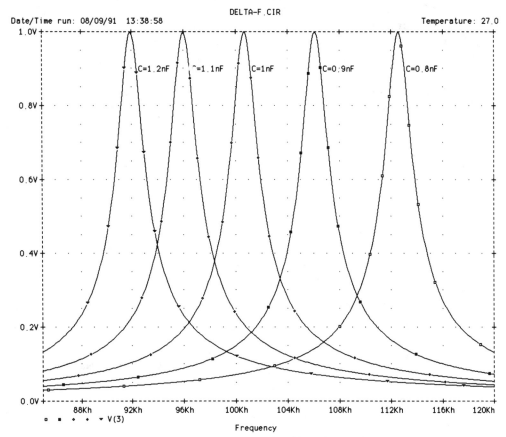

Figure 12.85 Resonant Circuit Resistor Voltage vs. Frequency.

the relationship between capacitor value and resonant frequency is not linear, shown by the uneven spacing between resonant peaks.

EXAMPLE 12.6.6, SENS.CIR

PSpice has the capability to determine the small-signal DC sensitivity of an output variable (node voltage or current through a voltage source) to all parameters in the circuit. Two instances in which this can be useful are when determining which parts in a circuit must be very close to their design values, and in finding parameters such as power supply rejection ratio in an amplifier circuit. Figure 12.86 shows a differential amplifier with two voltage source inputs.

Input file SENS.CIR in Fig. 12.87 includes the control line .SENS V(4), which instructs PSpice to print out the name and the sensitivities (measured both in V/unit and V/percent) of each element in the circuit which will affect the DC voltage at node 4. If the output variable in the .SENS line was a current, the sensitivity would be expressed in A/unit and A/percent.

An examination of the output file, Fig. 12.88, reveals the results of the sensitivity analysis. As an example of what

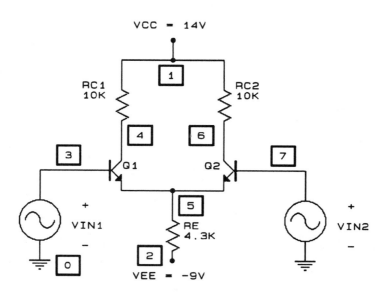

Figure 12.86 Differential Amplifier Circuit.

```
SENS.CIR    DIFF-AMP WITH SENSITIVITY ANALYSIS
VCC  1  0  14
VEE  2  0  -9
VIN1 3  0  AC  1M
Q1   4  3  5  MOD1
RC1  1  4  10K
RE   5  2  4.3K
VIN2 7  0  0
Q2   6  7  5  MOD1
RC2  1  6  10K
.SENS V(4)
.MODEL MOD1 NPN(BF=75)
.OPTIONS NOPAGE
.END
```

Figure 12.87 Input File SENS.CIR.

```
**** 08/09/91 13:58:13 ********* Evaluation PSpice (January 1991) ************
 SENS.CIR    DIFF-AMP WITH SENSITIVITY ANALYSIS
*****       CIRCUIT DESCRIPTION
*****************************************************************************
VCC  1  0  14
VEE  2  0  -9
VIN1 3  0  AC  1M
Q1   4  3  5  MOD1
RC1  1  4  10K
RE   5  2  4.3K
VIN2 7  0  0
Q2   6  7  5  MOD1
RC2  1  6  10K
.SENS V(4)
.MODEL MOD1 NPN(BF=75)
.OPTIONS NOPAGE
.END

    ****      BJT MODEL PARAMETERS
               MOD1
               NPN
          IS  100.000000E-18
          BF   75
          NF   1
          BR   1
          NR   1

  ****     SMALL SIGNAL BIAS SOLUTION       TEMPERATURE =   27.000 DEG C
  NODE   VOLTAGE     NODE   VOLTAGE     NODE   VOLTAGE     NODE   VOLTAGE
  (  1)  14.0000  (    2)  -9.0000  (   3)   0.0000  (    4)   4.5593
  (  5)  -.7727   (    6)   4.5593  (   7)   0.0000

      VOLTAGE SOURCE CURRENTS
      NAME          CURRENT
      VCC          -1.888E-03
      VEE           1.913E-03
      VIN1         -1.259E-05
      VIN2         -1.259E-05
      TOTAL POWER DISSIPATION   4.37E-02  WATTS
```

Figure 12.88 Output File SENS.OUT.

```
****        DC SENSITIVITY ANALYSIS              TEMPERATURE =    27.000 DEG C
DC SENSITIVITIES OF OUTPUT V(4)
            ELEMENT         ELEMENT             ELEMENT         NORMALIZED
            NAME            VALUE               SENSITIVITY     SENSITIVITY
                                                (VOLTS/UNIT)  (VOLTS/PERCENT)
            RC1             1.000E+04           -9.441E-04      -9.441E-02
            RE              4.300E+03            2.189E-03       9.411E-02
            RC2             1.000E+04           -4.644E-12      -4.644E-10
            VCC             1.400E+01            1.000E+00       1.400E-01
            VEE            -9.000E+00            1.144E+00      -1.030E-01
            VIN1            0.000E+00           -1.831E+02      -0.000E+00
            VIN2            0.000E+00            1.819E+02       0.000E+00
Q1
            RB              0.000E+00            0.000E+00       0.000E+00
            RC              0.000E+00            0.000E+00       0.000E+00
            RE              0.000E+00            0.000E+00       0.000E+00
            BF              7.500E+01           -8.255E-04      -6.192E-04
            ISE             0.000E+00            0.000E+00       0.000E+00
            BR              1.000E+00            1.000E-12       1.000E-14
            ISC             0.000E+00            0.000E+00       0.000E+00
            IS              1.000E-16           -4.735E+16      -4.735E-02
            NE              1.500E+00            0.000E+00       0.000E+00
            NC              2.000E+00            0.000E+00       0.000E+00
            IKF             0.000E+00            0.000E+00       0.000E+00
            IKR             0.000E+00            0.000E+00       0.000E+00
            VAF             0.000E+00            0.000E+00       0.000E+00
            VAR             0.000E+00            0.000E+00       0.000E+00
Q2
            RB              0.000E+00            0.000E+00       0.000E+00
            RC              0.000E+00            0.000E+00       0.000E+00
            RE              0.000E+00            0.000E+00       0.000E+00
            BF              7.500E+01           -8.255E-04      -6.192E-04
            ISE             0.000E+00            0.000E+00       0.000E+00
            BR              1.000E+00            4.919E-21       4.919E-23
            ISC             0.000E+00            0.000E+00       0.000E+00
            IS              1.000E-16            4.706E+16       4.706E-02
            NE              1.500E+00            0.000E+00       0.000E+00
            NC              2.000E+00            0.000E+00       0.000E+00
            IKF             0.000E+00            0.000E+00       0.000E+00
            IKR             0.000E+00            0.000E+00       0.000E+00
            VAF             0.000E+00            0.000E+00       0.000E+00
            VAR             0.000E+00            0.000E+00       0.000E+00

            JOB CONCLUDED
            TOTAL JOB TIME              2.08
```

Figure 12.88 (Continued).

the results mean, let's look at the effect of RE, the emitter resistor, on V(4). The sensitivity is 2.189E-03 V/unit, or 2.189 mV/ohm since RE has units of ohms. This means that if RE increases by one ohm, then V(4) will increase by 2.189 mV.

Since components frequently have tolerances expressed in percent, the RE sensitivity of 9.411E-02 V/percent is a useful result as well. If RE increases by 1%, or 43 ohms, then V(4) will increase by 94.11 mV. Note that 94.11 mV/2.189 mV = 43.

The sensitivity to VIN1 and VIN2 is -183.1 and 181.9, respectively. These are the inverting and non-inverting gains, respectively, of the differential amplifier. The fact that these gain magnitudes are slightly different accounts for the small, but non-zero, common-mode gain of a differential amplifier. The reason that the normalized sensitivity, in V/percent, is zero is that the DC values of VIN1 and VIN2 are both zero.

Significant insight can be obtained from the sensitivity analysis results. The sensitivity of V(4) to RC1 is 200 million times larger than the sensitivity to RC2, since

9.441E-02/4.644E-10 = 203E6

Also, significantly huge amounts of output data can be generated by using the .SENS control line in a large circuit. Be sure to examine the output file with a text editor before printing it and consuming massive amounts of paper.

EXAMPLE 12.6.7, NOISE.CIR

In Fig. 12.89, the last differential amplifier to appear in this chapter, a noise analysis will be done. Noise analysis can only be done along with an AC analysis. PSpice must be told across which output nodes the noise voltage is assumed to appear, and which input independent source (voltage or current) is to be used for the input noise reference.

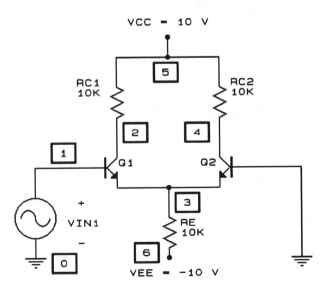

Figure 12.89 Differential Amplifier Circuit.

```
NOISE.CIR   DIFFERENTIAL AMPLIFIER WITH NOISE ANALYSIS
VIN1 1 0 AC 10M
Q1 2 1 3 MOD1
Q2 4 0 3 MOD1
.MODEL MOD1 NPN(RB=2  RC=.8  RE=.5  KF=1E-12)
VCC 5 0 10
VEE 6 0 -10
RC1 5 2 10K
RC2 5 4 10K
RE 3 6 10K
.AC DEC 1 1 100MEG
.NOISE V(2,3) VIN1   8
.OPTIONS NOPAGE
.PROBE
.END
```

Figure 12.90   Input File NOISE.CIR.

The input file, Fig. 12.90, instructs PSpice to perform an AC analysis from 1 Hz to 100 MHz with one frequency per decade, for a total of nine frequencies. The .NOISE control line specifies V(2,3) as the assumed output and VIN1 as the input noise reference, and also specifies that the summary printout of all parameters that contribute to noise (resistors and semiconductors) be printed at every eighth frequency.

Output file NOISE.OUT, Fig. 12.91A, contains a summary printout of the noise at 1 Hz and 100 MHz. All output noise contributions are the same at both frequencies except for the flicker noise (FN) of the transistors. The input noise multiplied by the voltage gain (transfer function value of 88.41 V/V) gives the output noise.

The Probe display in Fig. 12.91B is the graph of input noise and output noise versus frequency, which shows that the flicker noise is significant below 100 KHz.

```
**** 08/09/91 14:11:41 ********* Evaluation PSpice (January 1991) ************
 NOISE.CIR   DIFFERENTIAL AMPLIFIER WITH NOISE ANALYSIS
 ****      CIRCUIT DESCRIPTION
******************************************************************************
VIN1 1 0 AC 10M
Q1 2 1 3 MOD1
Q2 4 0 3 MOD1
.MODEL MOD1 NPN(RB=2  RC=.8  RE=.5  KF=1E-12)
VCC 5 0 10
VEE 6 0 -10
RC1 5 2 10K
RC2 5 4 10K
RE 3 6 10K
.AC DEC 1 1 100MEG
.NOISE V(2,3) VIN1   8
.OPTIONS NOPAGE
.PROBE
.END
```

Figure 12.91A   Output File NOISE.OUT.

```
****     BJT MODEL PARAMETERS
              MOD1
              NPN
         IS  100.000000E-18
         BF  100
         NF  1
         BR  1
         NR  1
         RB  2
        RBM  2
         RE  .5
         RC  .8
         KF  1.000000E-12


****     SMALL SIGNAL BIAS SOLUTION       TEMPERATURE =   27.000 DEG C
   NODE  VOLTAGE      NODE  VOLTAGE      NODE  VOLTAGE      NODE  VOLTAGE
(    1)   0.0000   (    2)   5.4229   (    3)   -.7542   (    4)   5.4229
(    5)  10.0000   (    6) -10.0000

     VOLTAGE SOURCE CURRENTS
     NAME          CURRENT
     VIN1         -4.577E-06
     VCC          -9.154E-04
     VEE           9.246E-04

     TOTAL POWER DISSIPATION   1.84E-02  WATTS


****     NOISE ANALYSIS                  TEMPERATURE =   27.000 DEG C
         FREQUENCY =  1.000E+00 HZ


**** TRANSISTOR SQUARED NOISE VOLTAGES (SQ V/HZ)
            Q1         Q2
RB       2.591E-16  2.505E-16
RC       0.000E+00  0.000E+00
RE       6.479E-17  6.262E-17
IB       3.859E-17  3.228E-17
IC       3.737E-15  3.471E-15
FN       1.204E-10  1.007E-10
TOTAL    1.204E-10  1.007E-10


**** RESISTOR SQUARED NOISE VOLTAGES (SQ V/HZ)
            RC1        RC2         RE
TOTAL    1.658E-16  0.000E+00  3.994E-17


**** TOTAL OUTPUT NOISE VOLTAGE          = 2.212E-10 SQ V/HZ
                                         = 1.487E-05 V/RT HZ

     TRANSFER FUNCTION VALUE:
       V(2,3)/VIN1                       = 8.841E+01
     EQUIVALENT INPUT NOISE AT VIN1 =  1.682E-07 V/RT HZ


****     NOISE ANALYSIS                  TEMPERATURE =   27.000 DEG C
         FREQUENCY =  1.000E+08 HZ
```

Figure 12.91A (Continued).

```
**** TRANSISTOR SQUARED NOISE VOLTAGES (SQ V/HZ)
            Q1          Q2
RB      2.591E-16   2.505E-16
RC      0.000E+00   0.000E+00
RE      6.479E-17   6.262E-17
IB      3.859E-17   3.228E-17
IC      3.737E-15   3.471E-15
FN      1.204E-18   1.007E-18
TOTAL   4.100E-15   3.818E-15

**** RESISTOR SQUARED NOISE VOLTAGES (SQ V/HZ)
            RC1         RC2         RE
TOTAL   1.658E-16   0.000E+00   3.994E-17

**** TOTAL OUTPUT NOISE VOLTAGE       = 8.124E-15 SQ V/HZ
                                      = 9.013E-08 V/RT HZ
          TRANSFER FUNCTION VALUE:
            V(2,3)/VIN1               = 8.841E+01
          EQUIVALENT INPUT NOISE AT VIN1 = 1.019E-09 V/RT HZ

          JOB CONCLUDED
          TOTAL JOB TIME            2.69
```

Figure 12.91A (Continued).

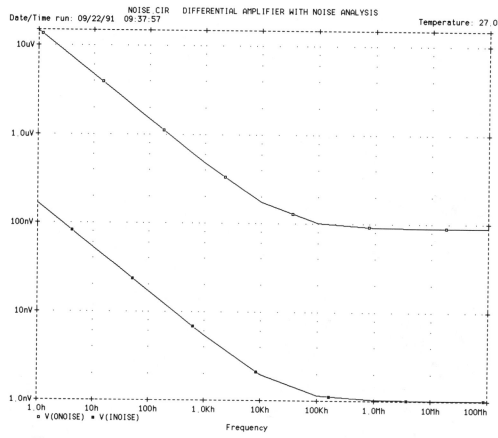

Figure 12.91B Display of Noise Voltage vs. Frequency Using Probe.

CHAPTER SUMMARY

A good reference to use when generating PSpice input files is a similar input file. 30 example files covering a wide variety of analysis types were presented in this chapter. Additional example files may be found in Chaps. 3, 9, 10 and 11.

Organize the input files you generate, along with your notes about details of element and control lines. These will be good references for future PSpice work you do.

The example circuits in this chapter, with brief descriptions of the PSpice functions they illustrate, follow:

Input File Name	Circuit Type & PSpice Functions Illustrated	
DCMESS.CIR	DC circuit	
VANDI.CIR	DC circuit	
DC-CKT.CIR	DC circuit with VCCS	
DI-VA.CIR	Diode V-A characteristic	.DC
FETINV.CIR	JFET inverter	.DC
TTLINV.CIR	7404 TTL logic gate	.DC
LOPASS.CIR	R-C filter	.AC
WEIN.CIR	Wein bridge	.AC
LCMATCH.CIR	Many L, C elements	.AC
SERIESAC.CIR	2 series resonant circuits	.AC
DUBTUNE.CIR	Double-tuned transformer	.AC
2AC.CIR	2-source problem	.AC
2HP.CIR	2 active filters in parallel	.AC
ACDIFAMP.CIR	BJT differential amp	.AC
TRIANG.CIR	PWL triangle, R-C circuit	.TRAN
LC.CIR	PWL current pulse into L-C	.TRAN
EKG-LP.CIR	PWL electro-cardiogram	.TRAN
PWRSUPLI.CIR	1/2-wave rectifier	.TRAN
CMOSNAND.CIR	MOSFET nand gate	.TRAN
TTL.CIR	BJT 7404 TTL inv. gate	.TRAN
ASTABLE.CIR	2-BJT clock, w/initial cond.	.TRAN,.IC, UIC
ABSVAL.CIR	Prec. rectifier, subcircuits	.TRAN
AM.CIR	Non-linear VCVS, ampl. mod.	.TRAN
VARYTEMP.CIR	Diode V-A graph, 2 temps.	.TEMP, .DC
DELTAT.CIR	BJT diff-amp, transfer func.	.TEMP, .TF
FOURIER.CIR	1/2-wave rect., Fourier anal.	.TRAN, .FOUR
LH0005.CIR	Op-amp, open-loop	.TF, .OP
DELTA-F.CIR	Resonant circuit, variable C	.AC, .STEP
SENS.CIR	BJT diff-amp, sensitivity	.SENS
NOISE.CIR	BJT diff-amp, noise analysis	.AC, .NOISE

ANSWERS TO PROBLEMS

Chapter 4

Problem 1.

```
CHAP4PR1.CIR    3 R'S, 2 BATTERIES
* SOLUTION TO CHAPTER 4, PROBLEM 1.
VL 0 5 10
RL 5 12 2K
RM 12 0 3K
RR 12 38 1K
VR 38 0 20
.OPTIONS NOPAGE
.END
```

Problem 2.

```
CHAP4PR2.CIR  PARALLEL RESONANT CIRCUIT, 10 KHZ
* SOLUTION TO CHAPTER 4, PROBLEM 2.
R1  1 2 1K
R2  3 0 1
VGEN 1 0 AC 1
L1  2 3 1M
C1  2 0 0.2533U
.AC LIN 21  5000 15000
.PRINT AC VM(2) VP(2)
.PLOT AC  VM(2) VP(2)
.PROBE
.END
```

234

Problem 3.

```
CHAP4PR3.CIR  R-L CIRCUIT WITH PULSE INPUT
* SOLUTION TO CHAPTER 4, PROBLEM 3
VIN 1 0 PULSE(0 1 10U 1N 1N 80U)
RA  1  2  1000
LA  2  0  20E-3
.TRAN 5U 160U
.PRINTTRAN V(2) V(1)
.PLOT TRAN V(2) V(1)
.PROBE
.END
```

Problem 4.

```
CHAP4PR4.CIR    TWO-SOURCE DC BRIDGE
* SOLUTION TO CHAPTER 4, PROBLEM 4.
V1  14  0  5
R1  14 16 10
R2  16  0 40
R3  14 18 20
R4  18 12 50
R5  16 18 30
V2  12  0  7
.OPTIONS NOPAGE
.END
```

Chapter 6

Problem 1.

a.

```
BRIDGE.CIR  DC BRIDGE CIRCUIT
* SOLUTION TO CHAPTER 6, PROBLEM 1.
V   10  0 20
RA  10 12  6K
RB  12 14  2K
RC  12 16  5K
RD  14 16  4K
RE  14  0  3K
RF  16  0  1K
.OP
.OPTIONS NOPAGE
.END
```

   c. $V(14) - V(16) = 3.0343$ V - 1.3377 V = 1.6966 V.

Problem 2.

a,c.

```
OH-MY.CIR   DC CIRCUIT WITH A LOT HAPPENING
* SOLUTION TO CHAPTER 6, PROBLEM 2.
VL  1 0 10
R1  1 2 1
I   0 2 1
```

```
R2  2  3  2
VSENSE  4  3  0
R4  2  5  5
E   4  0  2  1  7
R3  4  5  3
VR  5  0  3
.SENS I(VSENSE)
.OP
.OPTIONS NOPAGE
.END
```

b. I(VSENSE) = 4 A, from node 4 to node 3.

Problem 3.

a.

```
PRE-EMPH.CIR  A PRE-EMPHASIS CIRCUIT FOR FM
*  SOLUTION TO CHAPTER 6, PROBLEM 3.
VIN1  0  AC  1
RA  1  2  47K
CA  1  2  1.6E-9
RB  2  0  5K
.AC DEC 10 200 200K
.PLOT AC VM(2) (0.1,1)
.OPTIONS NOPAGE
.PROBE
.END
```

Problem 4.

a.

```
NOTCH.CIR   ACTIVE NOTCH FILTER, AC ANALYSIS
*  SOLUTION TO CHAPTER 6, PROBLEM 4A.
VIN1  0  AC  1
*    ACTIVE HIGH-PASS FILTER FOLLOWS
C1  1  2  1E-7
C2  2  3  1E-7
R1  3  0  2K
R2  2  5  2K
R3  5  4  818
R4  4  0  2K
*    FIRST OP-AMP MODEL FOLLOWS
RIN1  3  4  1MEG
E1  5  0  3  4  1E5
*    ACTIVE LOW-PASS FILTER FOLLOWS
R5  5  6  2K
R6  6  7  2K
R7  9  8  818
R8  8  0  2K
C3  9  6  0.1U
C4  7  0  0.1U
*    SECOND OP-AMP MODEL FOLLOWS
RIN2  7  8  1MEG
E2  9  0  7  8  1E5
*    TRUE DIFFERENTIAL AMPLIFIER FOLLOWS
R9  1  10  2K
R10 10  0  2K
R11 9  11  2K
R12 11  12  2K
```

```
*    THIRD OP-AMP MODEL FOLLOWS
RIN3  10  11  1MEG
E3  12  0  10  11  1E5
.AC DEC 15 100 10K
.PLOT AC  VDB(12)
.OPTIONS NOPAGE
.PROBE
.END
```

Problem 4.

b. Notch depth, or gain at 800 Hz., = -13.31 dB.

```
NOTCH2.CIR   ACTIVE NOTCH FILTER, AC ANALYSIS, NARROW BAND
* SOLUTION TO CHAPTER 6, PROBLEM 4B.
VIN1 0 AC 1
*    ACTIVE HIGH-PASS FILTER FOLLOWS
C1  1  2  1E-7
C2  2  3  1E-7
R1  3  0  2K
R2  2  5  2K
R3  5  4  818
R4  4  0  2K
*    FIRST OP-AMP MODEL FOLLOWS
RIN1  3  4  1MEG
E1  5  0  3  4  1E5
*    ACTIVE LOW-PASS FILTER FOLLOWS R55  6  2K
R6  6  7  2K
R7  9  8  818
R8  8  0  2K
C3  9  6  0.1U
C4  7  0  0.1U
*    SECOND OP-AMP MODEL FOLLOWS
RIN2  7  8  1MEG
E2  9  0  7  8  1E5
*    TRUE DIFFERENTIAL AMPLIFIER FOLLOWS
R9  1  10  2K
R10 10  0  2K
R11 9  11  2K
R12 11  12  2K
*    THIRD OP-AMP MODEL FOLLOWS
RIN3  10  11  1MEG
E3  12  0  10  11  1E5
.AC LIN 31  500  1100
.PLOT AC  VDB(12)
.OPTIONS NOPAGE
.PROBE
.END
```

Problem 5.

a.

```
MANY-C.CIR   7 CAPS, TO DETERMINE EFFECTIVE CAPACITANCE
* SOLUTION TO CHAPTER 6, PROBLEM 5A.
C1  1  3  2U
C2  1  2  3U
C3  2  3  1U
C4  2  4  1.5U
C5  3  4  2.5U
C6  3  0  3.5U
```

```
C7  4  0  0.5U
L   10 1  1MH
V   10 0  AC
*   DUMMY RESISTORS ARE NEEDED TO GIVE DC PATHS TO GROUND
R3  3  0  1G
R4  4  0  1G
R2  2  0  1G
.AC DEC 5 1 1MEG
.PLOT AC VM(1) VP(1)
.OPTIONS NOPAGE
.PROBE
.END
```

b.

```
MANY-C3.CIR  7 CAPS, TO DETERMINE EFFECTIVE CAPACITANCE
*   SOLUTION TO CHAPTER 6, PROBLEM 5B.
*   NARROWER FREQUENCY RANGE
C1  1  3  2U
C2  1  2  3U
C3  2  3  1U
C4  2  4  1.5U
C5  3  4  2.5U
C6  3  0  3.5U
C7  4  0  0.5U
L   10 1  1MH
V   10 0  AC
*   DUMMY RESISTORS ARE NEEDED TO GIVE DC PATHS TO GROUND
R3  3  0  1G
R4  4  0  1G
R2  2  0  1G
.AC  LIN  26  3.7K  3.8K
.PLOT AC VM(1) VP(1)
.OPTIONS NOPAGE
.PROBE
.END
```

Problem 6.

a.

```
HP-RING.CIR  HIGH-PASS ACTIVE FILTER, UNDER-DAMPED, RINGS
*   SOLUTION TO CHAPTER 6, PROBLEM 6.
VIN1 0 PULSE(0 1 0.5M 1U 1U 0.4M)
C1  1  2  8N
C2  2  3  8N
R1  5  2  2K
R2  3  0  2K
R3  4  0  2K
R4  5  4  3.96K
*   OP-AMP MODEL FOLLOWS
RIN3 4 1E6
E   5  0  3  4  5E4
.TRAN 0.1M 5M
.PLOT TRAN V(5) V(1)
.OPTIONS NOPAGE
.PROBE
.END
```

Problem 7.

a.

```
SQ-WAVE.CIR   FOURIER ANALYSIS OF A 1 KHZ SQUAREWAVE
*  SOLUTION TO CHAPTER 6, PROBLEM 7.
V   1 0 PULSE(-1 1 0 1N 1N 0.5M 1M)
R   1 0 1
.TRAN 0.01M 2M
.FOUR 1K V(1)
.OPTIONS NOPAGE
.END
```

Appendix A

OTHER CONTROL LINES

A.1   INTRODUCTION

Seven control lines were not covered previously in this book, since they are the kinds of control lines you can generally do without until you start doing some serious and advanced PSPice analysis.   The control lines are: .WIDTH, .OPTIONS, .NODESET, .IC, .LIB, .MODEL and .STEP.   Browse through this appendix, noting particularly the many choices that can be specified using the .OPTIONS control line.   Try those that you wish to know more about, and compare the output file of the analysis with and without the options specified.

A.2   .WIDTH CONTROL LINE

The general form for the .WIDTH control line is

.WIDTH OUT = COLNUMOUT

where COLNUMOUT is the width of the output file; allowable values are 80 and 132.   The default used by PSpice in creating output files is 80 columns.   When the output file is viewed on a monitor, lines wider than 80 columns will generally wrap around, making the output file hard to interpret. Ordinarily it is best not to specify a column width of 132.

EXAMPLE:

.WIDTH OUT = 132

The output file will have a width of 132 columns.

A.3 .OPTIONS CONTROL LINE

The general form for the .OPTIONS control line is

.OPTIONS OPT1 OPT2 . . . (or OPT = OPTVAL . . .)

There are 32 options which may be specified; normally only a few are used in any particular input file. Options can be listed in any order. Many SPICE users will have little need to modify the default option values. For more information on these options, refer to the publications in Appendix D. In the listing below, "x" stands for a positive number. The options and their effects are:

Option	Effect
ACCT	causes accounting and run time statistics to be printed.
LIST	causes the summary listing of the input data to be printed.
NOMOD	suppresses the printout of the model parameters.
NOPAGE	suppresses page ejects.
NODE	causes the printing of the node table.
OPTS	causes the option values to be printed.
GMIN=x	resets the value of GMIN, the minimum conductance allowed by the program. The default value is 1.0E-12.
RELTOL=x	resets the relative error tolerance of the program. The default value is 0.001 (0.1 percent).
ABSTOL=x	resets the absolute current error tolerance of the program. The default value is 1 picoamp.
VNTOL=x	resets the absolute voltage error tolerance of the program. The default value is 1 microvolt.
TRTOL=x	resets the transient error tolerance. The default value is 7.0. This parameter is an estimate of

	the factor by which SPICE overestimates the actual truncation error.
CHGTOL=x	resets the charge tolerance of the program. The default value is 1.0E-14.
PIVTOL=x	resets the absolute minimum value for a matrix entry to be accepted as a pivot. The default value is 1.0E-13.
PIVREL=x	resets the relative ratio between the largest column entry and an acceptable pivot value. The default value is 1.0E-3. In the numerical pivoting algorithm the allowed minimum pivot value is determined by: EPSREL = Amax1(PIVREL*Maxval,PIVTOL) where Maxval is the maximum element in the column where a pivot is sought (partial pivoting).
NUMDGT=x	resets the number of significant digits printed for output variable values. x must satisfy the relation $0 < x < 8$. The default value is 4. Note: this option is independent of the error tolerance used by SPICE (i.e., if the values of options RELTOL, ABSTOL, etc., are not changed then one may be printing numerical noise for NUMDGT > 4).
TNOM=x	resets the nominal temperature. The default value is 27 deg C (300 deg K).
ITL1=x	resets the DC iteration limit. The default is 100.
ITL2=x	resets the DC transfer curve iteration limit. The default is 50.
ITL3=x	resets the lower transient analysis iteration limit. The default value is 4.
ITL4=x	resets the transient analysis timepoint iteration limit. The default is 10.
ITL5=x	resets the transient analysis total iteration limit. The default is 5000. Set ITL5 = 0 to omit this test.
CPTIME=x	the maximum cpu-time in seconds allowed for this job.
LIMTIM=x	resets the amount of cpu time reserved by SPICE for generating plots should a cpu timelimit cause job termination. The default value is 2 (sec.).
LIMPTS=x	resets the total number of points that can be printed or plotted in a DC, AC or transient analysis. The default value is 201.
LVLCOD=x	if x is 2 (two), then machine code for the matrix solution will be generated. Otherwise, no ma-

	chine code is generated. The default value is 2. Applies only to CDC computers.
LVLTIM=x	if x is 1 (one), the iteration timestep control is used. If x is 2 (two), the truncation-error timestep is used. The default value is 2. If method=Gear and MAXORD>2 then LVLTIM is set to 2 by SPICE.
Method=name	sets the numerical integration method used by SPICE. Possible names are Gear or trapezoidal. The default is trapezoidal.
MAXORD=x	sets the maximum order for the integration method if Gear's variable-order method is used. x must be between 2 and 6. The default value is 2.
DEFL=x	resets the value for MOS channel length; the default is 100.0 micrometer.
DEFW=x	resets the value for MOS channel width; the default is 100.0 micrometer.
DEFAD=x	resets the value for MOS drain diffusion area; the default is 0.0.
DEFAS=x	resets the value for MOS source diffusion area; the default is 0.0.

EXAMPLE:

.OPTIONS NOPAGE NODE LIMPTS = 501

This control line will shorten the output file by suppressing page ejects, will print a node table, and will increase from 201 (the default value) to 501 the maximum number of points that can be plotted or printed in an AC, DC or transient analysis.

EXAMPLE:

.OPTIONS TNOM = 0 ACCT RELTOL = .01 ITL5 = 0

This control line changes the nominal temperature to 0 degrees C, causes the accounting and run-time statistics to be printed, resets the relative error tolerance of the program from .001 (the default value) to .01 (1 percent), and omits the total iteration limit in a transient analysis. Using ITL5=0 in a long transient analysis will prevent PSpice from aborting before the job is finished.

A.4 .NODESET CONTROL LINE

The general form for the .NODESET control line is

.NODESET V(NODENUM1) = VAL1 V(NODENUM2) = VAL2 . . .

where NODENUM is a positive integer representing a node other than node 0, and VAL is the voltage at that node. PSpice will do the initial attempt to find a DC or transient bias point solution with the nodes specified held to the values shown. It is not necessary to specify the voltages at all nodes. Subsequently this restriction is released and PSpice continues to the correct solution.

The .NODESET line may be of use in bi-stable or astable circuits, where the program user can make an educated guess as to certain node voltages.

EXAMPLE:

.NODESET V(12) = -5 V(8) = 12

The voltage at node 12 is set to -5 V, and the node 8 voltage is set to 12 V for the initial bias point solution.

A.5 .IC CONTROL LINE

The general form for the .IC control line is

.IC V(NODENUM1) = VAL1 V(NODENUM2) = VAL2 . . .

where NODENUM is a positive integer representing a node other than node 0, and VAL is the voltage at that node. This line is used to set transient initial conditions. It has two different interpretations, depending on whether the UIC parameter is specified on the .TRAN control line. Also, one should not confuse this line with the .NODESET line (see A.4 above). The .NODESET line is only to help DC convergence, and does not affect final bias solution (except for multistable circuits). The two interpretations of this line are as follows:

1. When the UIC parameter is specified on the .TRAN line, then the node voltages specified on the .IC line are used to compute the capacitor, diode, BJT, JFET and MOSFET initial conditions. This is equivalent to specifying the IC = . . . parameter on each device element line, but is

much more convenient. The IC = . . . parameter can still be specified and will take precedence over the .IC values. Since no DC bias (initial transient) solution is computed before the transient analysis, one should take care to specify all DC source voltages on the .IC line if they are to be used to compute device initial conditions.

2. When the UIC parameter is not specified on the .TRAN line, the DC bias (initial transient) solution will be computed before the transient analysis. In this case, the node voltages specified on the .IC line will be forced to the desired initial values during the bias solution. During transient analysis, the constraint on these node voltages is removed.

EXAMPLE:

.IC V(12) = -5 V(8) = 12

The voltage at node 12 is set to -5 V, and the node 8 voltage is set to 12 V for the final bias point solution if the UIC parameter is not specified on the .TRAN control line. If the UIC parameter is specified, then the voltages for nodes 12 and 8 are used to compute the capacitor, diode, BJT, JFET and MOSFET initial conditions.

A.6 .LIB CONTROL LINE

The general form for the .LIB control line in a PSpice input file is

.LIB "FILENAME.LIB"

where FILENAME.LIB is a file in the PSpice subdirectory which contains a .MODEL line for a device in the PSpice input file.

EXAMPLE:

.LIB "EVAL.LIB"

In the PSpice input file a device is being used for which a model exists in the file EVAL.LIB, which comes with the PSpice evaluation version. This file includes diodes, bipolar transistors, JFETs, MOSFETs and other devices; it is a reduced version of the standard parts libraries available with the production version of PSpice.

EXAMPLE:

Q1 1 2 0 Q2N1893
.LIB "BANZHAF.LIB"

I made up a custom file of models for a variety of devices that I often use. The file, called BANZHAF.LIB, includes the model for a 2N1893 bipolar transistor and is in my PSpice sub-directory. If PSpice does not find the .MODEL line(s) for the 2N1893 in BANZHAF.LIB, an error message will be printed in the PSpice output file.

A.7 .MODEL CONTROL LINE

The .MODEL line has already been presented for switches in Section 2.6, and for semiconductors in Chap. 8. Here it is shown for resistors, capacitors and inductors. When .STEP is covered in the next section, the utility of .MODEL will be increased.

The general form for the .MODEL control line in a PSpice input file is

.MODEL MODNAME TYPE<(PAR1=PVAL1 PAR2=PVAL2 ...)>

where MODNAME is the model name given to a device in a separate element line, and TYPE tells PSpice what kind of device is being modeled. PAR is the parameter name, which can be any of those in the parameter list below, and PVAL is the value of that parameter. Note that the PAR and PVAL parts of the .MODEL line are optional; PSpice will use the default parameters for the device unless told otherwise.

TYPE	DEVICE NAME	PARAMETER NAMES
RES	Resistor	R - resistance multiplier
		TC1 - linear temperature coefficient
		TC2 - quadratic temperature coefficient
		TCE - exponential temp. coefficient
CAP	Capacitor	C - capacitance multiplier
		VC1 - linear voltage coefficient
		VC2 - quadratic voltage coefficient
		TC1 - linear temperature coefficient
		TC2 - quadratic temperature coefficient

IND	Inductor	L - inductance multiplier
		IL1 - linear current coefficient
		IL2 - quadratic current coefficient
		TC1 - linear temperature coefficient
		TC2 - quadratic temperature coefficient

EXAMPLE:

R1 7 8 DELTA 100K
.MODEL DELTA RES(TC1 = -.003)

Resistor R1 has a nominal value of 100K ohms at 27 degrees C. As temperature rises, R1 will drop by 0.3% per degree C.

EXAMPLE:

COUT 4 6 DRIFTY 10U
.MODEL DRIFTY CAP(VC1 = .02)

Capacitor COUT has a nominal value of 10 uF. For each volt across COUT, its capacitance will increase by 2% per volt.

A.8 .STEP CONTROL LINE

A quick way to see the effect on circuit behavior of a change in the value of a circuit element is to change it over a range and observe the analysis results. This can be very tedious when the changes are done manually by editing the input file to alter the element's value, and running a new analysis after each change.

The .STEP control line allows us to change one or more components automatically and see all the analysis results in a single Probe display. Here, the use of .STEP is limited to changing resistor, capacitor and inductor values. For more information on using .STEP, refer to the PSpice User's Guide.

The general form for the .STEP control line in a PSpice input file is

.STEP TYPE MODNAME(Prefix) Startval Stopval Incrval

where the TYPE is RES or CAP or IND, MODNAME is the model name given to the circuit element, and Prefix is the first letter of the element name (R for resistor, C for capacitor and L for inductor). Startval is the starting value, Stopval is the ending value, and Incrval is the increment size.

EXAMPLE:

COUT 4 6 DRIFTY 10U
.MODEL DRIFTY CAP
.STEP CAP DRIFTY(C) .6 1.5 .3

Capacitor COUT has a nominal value of 10 uF. It will be set to 0.6 of its nominal value, or 6 uF, and the PSpice analysis will be run. Then it will be incremented to 0.9 of its nominal value, or 9 uF, and the analysis will repeat. Thus, a total of four analyses will be done, with the capacitor value set to 6, 9, 12 and 15 uF.

EXAMPLE:

R1 7 8 DELTA 100K
.MODEL DELTA RES
.STEP RES DELTA(R) .95 1.05 .01

Resistor R1 has a nominal value of 100K ohms. It will be stepped from 95K ohm to 105K ohm in 1K ohm steps, with an analysis done at each step. Thus, a total of 11 analyses will be performed.

A.8.1 Example of Using .MODEL and .STEP

Figure A.1 shows a 1 uF capacitor being charged by a 1 mA current source. This should produce a ramp of voltage whose slope depends on the capacitor value. To see the effect of changing the capacitor, a transient analysis will be done and the capacitor will be stepped through a range of values. The

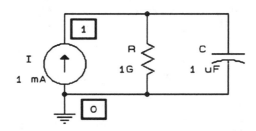

Figure A.1 Constant Current Into a Variable Capacitor.

```
RAMP-CAP.CIR  The value of a capacitor is stepped
I   0   1   DC   1E-3
R   1   0   1G
C   1   0   VARIABLE   1U   IC=0
.MODEL   VARIABLE   CAP
.STEP   CAP   VARIABLE(C)   .5   1.5   .5
.TRAN   .1M   10M   UIC
.PROBE
.END
```

Figure A.2 Input File RAMP-CAP.CIR.

1 gigaohm resistor is to provide a DC path to ground for node 1, without which PSpice will not work. Current sources and capacitors are considered to be open circuits during the small-signal bias solution which must precede a transient analysis.

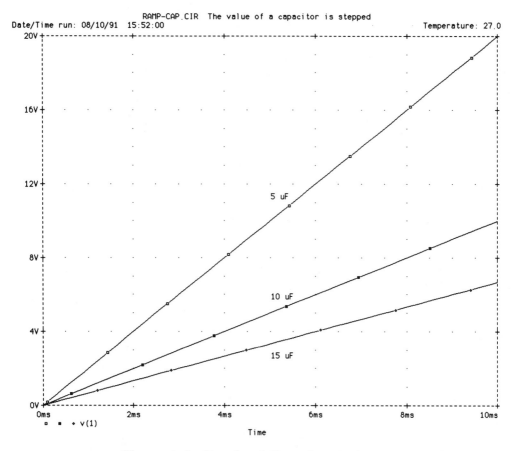

Figure A.3 Graph of Capacitor Voltage vs. Time for Three Capacitor Values.

The PSpice input file, Fig. A.2, sets the initial condition of the capacitor to 0 V at the start of each transient analysis. The .STEP . . . control line makes the capacitor be 0.5 uF, then 1 uF, and finally 1.5 uF. The analysis results, Fig. A.3, show that the rate of change of capacitor voltage is inversely proportional to the capacitance.

Appendix B

TRANSFORMERS
AND PSpice SWITCHES

B.1 ADVANCED HANDY HINTS

If you have gotten to this point in the book, your appetite for shortcuts, hints and neat ways of doing things with PSpice no doubt is whetted. In this appendix we will cover how to make an ideal transformer, how to make a lossless transformer using inductors, the use of voltage-controlled switches, and lastly how to make a switched three-phase power source.

No doubt as you use PSpice more and more you will encounter a circuit component which you would like to include in a PSpice analysis but which is not known to PSpice. Examples of such a device might be a mechanical relay, a varistor or a unijunction transistor. By spending some time looking into how devices are modeled, and learning how to use to their fullest capabilities those devices which PSpice does recognize, you will be able to create excellent approximations for most devices.

B.2 IDEAL TRANSFORMER, NO INDUCTORS

The ideal transformer is a fictitious device which divides voltage by a constant called turns ratio, multiplies current by the same ratio and changes impedance connected to it by the square of the turns ratio. Turns ratio is the number of primary turns divided by the number of secondary turns. Real transformers are magnetic machines with inductance, ohmic loss resistance and sometimes substantial losses in the magnetic core. An ideal transformer can be a useful circuit element to use when the inductance and losses of a real transformer can be neglected.

Figure B.1 shows the schematic symbol for an ideal transformer and the PSpice schematic for its model. It uses two controlled sources. The first is a voltage-controlled voltage source (VCVS), EPRI-SEC, which makes the secondary voltage equal to the product of the primary voltage and 1/(turns ratio). The second is a current-controlled current source (CCCS), FSEC-PRI, which makes the primary current equal to the secondary current multiplied by 1/(turns ratio).

In order to control FSEC-PRI a way of sensing the current flow in the secondary is needed. That is done by V-ISEC, a dead voltage source acting as an ammeter, which measures the current flowing up through EPRI-SEC.

N is the "gain" of the controlled sources, determined by the number of secondary winding turns divided by the number of primary winding turns

$$N = \# \text{ secondary turns}/(\# \text{ primary turns}) = 1/(\text{turns ratio})$$

The element lines below describe the ideal transformer model with N having a value of 2. In other words, the transformer is a 2:1 step-up transformer if the primary winding connections are nodes 10 and 11.

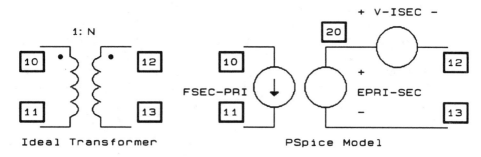

Figure B.1 Ideal Transformer and PSpice Model.

```
EPRI-SEC   20   13   10   11   2
FSEC-PRI   10   11   V-ISEC    2
V-ISEC     20   12   0
```

One word of caution: this PSpice circuit for an ideal transformer is a model based on behavior, not physics. It will work with DC sources as well as it will with AC sources. Real transformers often smoke and burn up when they are connected to DC sources.

B.3 IDEAL TRANSFORMER, WITH INDUCTORS

Another way to model a lossless transformer is to use two inductors which have a coefficient of coupling of 1.0. If the inductance of either the primary (LPRI) or the secondary (LSEC) is known, the inductance of the other winding can be determined using the number of primary and secondary turns, as follows:

N = # secondary turns/(# primary turns)

$LSEC = LPRI * N^2$, or $LPRI = LSEC/(N^2)$

The transformer and the PSpice model, assuming an N of 5 and primary inductance of 500 mH, is shown in Fig. B.2.

The secondary inductance can be calculated by

$LSEC = 50$ mH $* 5^2 = (5E-2)25 = 1.25$ H

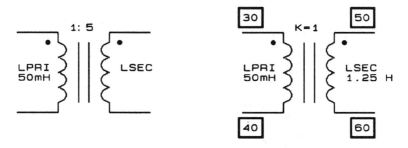

Figure B.2 Transformer and PSpice Model.

The element lines of this transformer model for use in a PSpice input file would be

```
LPRI      30   40   50E-3
LSEC      50   60   1.25
KPRI-SEC  LPRI LSEC    1.00
```

Notice that the first node for each inductor is the node which is connected to the dotted end in the schematic, Fig. B.2. This ensures the proper polarity of induced voltage in the secondary. Unlike the ideal transformer without inductance in B.2, this model with inductance will not work with DC. With DC current flowing through the primary, there is no changing magnetic flux in the core of the transformer, so there is no induced voltage in the secondary.

B.4 USING PSpice VOLTAGE-CONTROLLED SWITCHES

PSpice has both voltage-controlled and current-controlled switch models available. While using them in circuit modeling is rather straightforward, here's an example to help clarify the details of doing it.

A classic textbook problem in a first circuits course is illustrated in Fig. B.3. Initially, switch 1 (S1) is open and

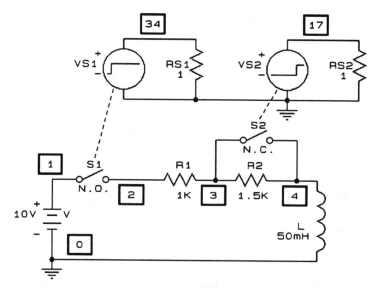

Figure B.3 L-R Charging Circuit with Switches.

switch 2 (S2) is closed. At 50 microseconds, S1 closes and at 300 microseconds S2 opens. The assignment is to sketch the current through the inductor, and the voltage across the inductor, for the first 400 microseconds. Assuming you know how to do this without PSpice already, here's how to model the switches.

Figure B.4, the PSpice input file, reveals that switch S1 is voltage-controlled, operated by the voltage between node 34 and node 0. Its .MODEL line tells us that when V(34,0) is less than 2 V, the switch is off (open), and when V(34,0) is above 3 V, the switch is on (closed). Thus, it is labeled as N.O., or normally open. Both its on and off resistances are negligible compared to the circuit impedances. The voltage controlling S1 is VS1, a piecewise linear voltage which steps from 0 V to 5 V between 49 and 50 microseconds. RS1 is there to satisfy the PSpice requirement that each node must have at least two circuit elements connected to it.

In a similar fashion, voltage-controlled switch S2 is controlled by VS2, which steps from 0 V to 5 V at 300 microseconds. However, note that S2 is on (closed) when V(17,0) is

```
SWITCH-L.CIR      Inductor charging through two resistors
*
************THE VOLTAGE CONTROLLING SWITCH S1 FOLLOWS
VS1  34  0  PWL(0 0, 49U 0, 50U 5)
*        vs1 is a step from 0 V to 5V at 50 usec
RS1  34  0  1
*
************NORMALLY OPEN VOLTAGE-CONTROLLED SWITCH S1 FOLLOWS
S1  1  2  34  0 EASY
.MODEL  EASY  VSWITCH  ron=.1  roff=1E9  von=3  voff=2
*
************THE VOLTAGE CONTROLLING SWITCH S2 FOLLOWS
VS2  17  0  PWL(0 0, 299U 0, 300U 5)
*        vcontrol is a step from 0V to 5V at 300 usec
RS2  17  0  1
*
***********NORMALLY CLOSED VOLTAGE-CONTROLLED SWITCH S2 FOLLOWS
S2  3  4  17  0 SIMPLE
.MODEL  SIMPLE  VSWITCH  ron=.01  roff=1E6  von=2  voff=3
*
***********L, 2Rs AND A BATTERY FOLLOW
V  1  0  DC  10
R1  2  3  1K
R2  3  4  1.5K
L  4  0  50M  IC=0
*        here are the control lines
.TRAN  1U  400U  0  1U  UIC
.PROBE
.END
```

Figure B.4 Input File SWITCH-L.CIR.

less than 2 V, and is off (open) when V(17,0) is above 3 V. For this reason, S2 is labeled as N.C., or normally closed.

In the input file inductor L is given the initial condition that its current at time zero is 0 A (IC=0), and the .TRAN control line ends with UIC (Use Initial Conditions). Without these specifications, the analysis is very incorrect. Figure B.5, the Probe display of analysis results, shows the inductor current in the upper plot and the inductor voltage in the lower plot. For both parts of the display (charging the inductor to 10 mA and discharging it to 4 mA), steady-state conditions are reached.

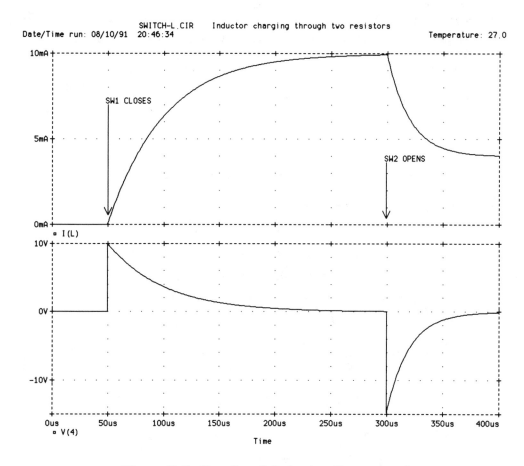

Figure B.5 Graphs of Inductor Current and Voltage vs Time.

B.5 USING PSpice CURRENT-CONTROLLED SWITCHES

For certain simulations involving three-phase power sources, it may be desirable to connect the power source to a three-phase load at any arbitrary time in the 360 degree input cycle. Load application might occur when phase A is at 0, 90 or 270 degrees (the three phases will be called A, B and C). In this section the generation of three sinusoidal voltages separated by 120 degrees of phase and the simultaneous switching of all three voltages to a load at an arbitrary point in the cycle will be illustrated.

The obvious way to generate a sinusoid is to use the independent voltage source SIN function, which was introduced in Section 2.5.1.2. The SIN function allows for any offset voltage, amplitude, frequency, delay time, damping factor or phase to be specified. Three sinusoidal voltage sources, identical except for the initial phase angle, will be used to make a WYE power source. To make it interesting, they will be connected simultaneously to a WYE load when the phase A voltage has reached 90 degrees.

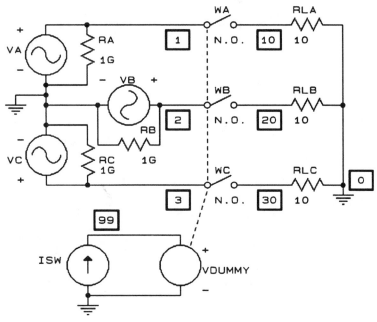

Figure B.6 Switched Three-Phase Power Circuit.

Figure B.6 is the three-phase circuit, with the three phases, sources VA, VB and VC, on the left. Across each a 1G ohm resistor is present to keep PSpice happy (the two-elements-to-a-node requirement rears its head again). WA, WB and WC are normally-open current-controlled switches, whose state (on or off) depends on the current through VDUMMY; notice that ISW, the current source whose current is sensed by VDUMMY, controls all three switches.

Input file 3-PHASE.CIR, Fig. B.7, makes ISW step from 0 A to 1 A at 4.1666 msec, which is 90 degrees of a 60 Hz sinusoid period. In this way, all three switches will close when phase A is at 90 degrees. Since the switches are identical, a single .MODEL line is used for all three. The 170 V magnitude is the peak voltage of a 120 VRMS voltage.

In the Probe display, Fig. B.8, VA by itself is shown in the upper plot and all three load voltages are in the lower plot. It can be seen that all phases are switched on when VA is at 90 degrees (4.1666 ms). By varying the delay time of ISW, the switching can be made to occur anywhere in the cycle.

```
3-PHASE.CIR    GENERATING SWITCHED 3-PHASE POWER
*
*          THE THREE FREE-RUNNING PHASES ARE CREATED BELOW
*
VA       1  0  SIN(0 170 60 0 0 0)
VB       2  0  SIN(0 170 60 0 0 120)
VC       3  0  SIN(0 170 60 0 0 240)
RA       1  0  1G
RB       2  0  1G
RC       3  0  1G
*
*          THE SWITCH CONTROL CURRENT FOR ALL 3 PHASES FOLLOWS
*
ISW      0  99  PULSE(0 1 4.1666M 1U)
VDUMMY   99 0  DC  0
*
*          THREE CURRENT-CONTROLLED SWITCHES FOLLOW
*
WA       1  10  VDUMMY   HIPOWER
WB       2  20  VDUMMY   HIPOWER
WC       3  30  VDUMMY   HIPOWER
.MODEL   HIPOWER  ISWITCH  ron=.1  roff=1g  ion=.8  ioff=.6
*          notice that only one .MODEL is used for 3 switches
RLA      10 0  10
RLB      20 0  10
RLC      30 0  10
.TRAN  0.1M   35M   0   .1M
.PROBE
.END
```

Figure B.7 Input File 3-PHASE.CIR.

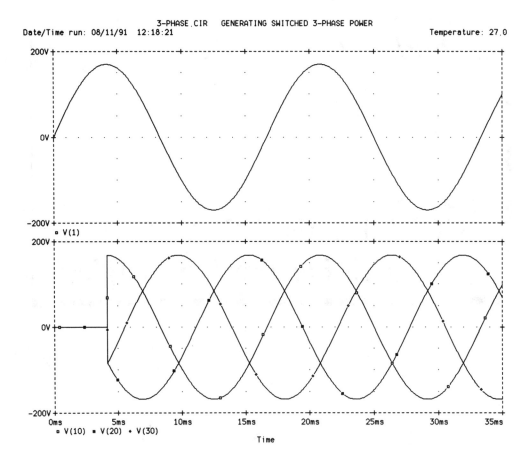

Figure B.8 Phase A Voltage and Three-Phase Load Voltages.

Appendix C

NONLINEAR (POLYNOMIAL) DEPENDENT SOURCES

C.1 NONLINEAR DEPENDENT SOURCES

In Chap. 4 the topic of linear dependent sources was covered. PSpice also supports nonlinear dependent sources, which have great utility for modeling nonlinear resistors, multipliers (including amplitude modulation generators) and voltage-controlled and current-controlled switches, to name a few. The usefulness of nonlinear dependent sources is limited by one's imagination (and patience with mastering the topic).

PSpice recognizes the same four types of dependent sources as mentioned in Chap. 5:

voltage-controlled current source, or VCCS
voltage-controlled voltage source, or VCVS
current-controlled current source, or CCCS
current-controlled voltage source, or CCVS.

The sources are uniquely specified by an element name beginning with the letter G, E, F or H and have units as described below:

ELEMENT	EQUATION	ELEMENT TYPE	UNIT
VCCS	I = G (V)	Transconductance	Siemen
VCVS	V = E (V)	Voltage gain	V/V
CCCS	I = F (I)	Current gain	A/A
CCVS	V = H (I)	Transresistance	Ohm

A nonlinear dependent source can, depending on how it is specified, be very linear. Perhaps a better name would be a polynomial source, since the equations that determine the source output are described by one or more polynomial coefficients. The set of coefficients is p0, p1, . . . , pn. The meaning of a coefficient depends on how many controlling sources there are upon which the polynomial source depends. There is no limit on the number of controlling sources except one's own ability to keep track of the coefficients. My limit is three, except on Mondays when it is two.

C.1.1 One-Dimensional Polynomial Equation

If a polynomial source is itself a function of only one controlling source, it is said to be one-dimensional. An expression for the polynomial source, called fv (for function value) in terms of the controlling source, called fa, is

$$fv = p0 + p1*fa + p2*fa^2 + p3*fa^3 + p4*fa^4 + \ldots$$

If this were a VCVS, then the unit of p0 would be volt, the unit of p1 would be volt/volt, the unit of p2 would be volt/(volt*volt), etc. In order to make it easy to write element lines for one-dimensional polynomial sources, if one and only one coefficient is specified PSpice will interpret it to be the p1 coefficient, and assumes p0 = 0.0.

C.1.2 Two-Dimensional Polynomial Equation

If a polynomial source is itself a function of two controlling sources, it is said to be two-dimensional. An expression for the polynomial source, called fv (for function value) in terms of the controlling sources, called fa and fb, is

$$fv = p0 + p1*fa + p2*fb + p3*fa^2 + p4*fa*fb + p5*fb^2$$
$$p6*fa^3 + p7*fa^2*fb + p8*fa*fb^2 + p9fb^3 + \ldots$$

If this were a VCCS, then the unit of p0 would be amp, the unit of p1 and p2 would be amp/volt, the unit of p3, p4 and p5 would be amp/(volt*volt), and the unit of p6, p7, p8 and p9 would be amp/(volt*volt*volt).

C.1.3   Three-Dimensional Polynomial Equation

If a polynomial source is itself a function of three controlling sources, it is said to be three-dimensional.   An expression for the polynomial source, called fv (for function value), in terms of the controlling sources, called fa, fb and fc, is

$$fv = p0 + p1*fa + p2*fb + p3*fc + p4*fa^2 + p5*fa*fb +$$
$$p6*fa*fc + p7*fb^2 + p8*fb*fc + p9*fc^2 + p10*fa^3 +$$
$$p11*fa^2*fb + p12*fa^2*fc + p13*fa*fb^2 + p14*fa*fb*fc$$
$$+ p15*fa*fc^2 + p16*fb^3 + p17*fb^2*fc + p18*fb*fc^2$$
$$+ p19*fc^3 + p20*fa^4 + \ldots$$

If this were a CCCS, then the unit of p0 would be amp, the unit of p1, p2 and p3 would be amp/amp, the unit of p4 through p9 would be amp/(amp*amp), and the unit of p10 through p19 would be amp/(amp*amp*amp).
Figuring out the coefficients for a four-dimensional polynomial source can be daunting.   If all you are doing is adding four controlling sources, then p0 = 0 and p1, p2, p3 and p4 would be 1. Beyond that, proceed slowly and cautiously.

C.2   NONLINEAR VOLTAGE-CONTROLLED CURRENT SOURCES

The general form for a nonlinear VCCS is

```
GXXXXXX  N+ N- POLY(ND) NC1+ NC1- <NC2+ NC2-> ...
+  P0  P1  ... <IC = ...>
```

where N+ and N- are the nodes to which the current source is connected. ND is the number of devices controlling the VCCS, and must be specified. NC+ and NC- are the positive and negative controlling nodes, respectively, and there must be a pair for each dimension. The initial condition, IC = , is the value of the controlling voltages at time zero. If initial conditions are not specified the default value is zero.

EXAMPLE:

GSIMPLE 1 4 POLY(1) 7 2 0 0.5 0.3 .06

The controlled current that flows from node 1 to node 4 through GSIMPLE is a function of the voltage at node 7 compared to node 2, and has the following equation:

$$I = 0 + 0.5*V(7,2) + 0.3*V(7,2)^2 + (6E\text{-}2)*V(7,2)^3 \text{ amp}$$

EXAMPLE:

GGOLLY 5 9 POLY(2) 1 0 3 4 .001 7M 80U

Voltage-controlled current source GGOLLY is connected to nodes 5 and 9 (with positive current assumed to flow from 5 to 9 through GGOLLY) and is controlled by two voltages. The first controlling voltage is the node 1 to node 0 voltage, the second is the voltage at node 3 compared to node 4. P0 = 1E-3, P1 = 7E-3 and P2 = 8E-5. This means that the expression for the current is

$$I = 1E\text{-}3 + (7E\text{-}3)*V(1,0) + (8E\text{-}5)*V(3,4) \text{ amp}$$

EXAMPLE:

It is possible to model a nonlinear resistor by using a nonlinear voltage-controlled current source of one dimension. Consider a VCCS whose controlling nodes are the same as the nodes between which the current flows. The current is then a nonlinear function of the voltage; in other words, a nonlinear resistor. Examine the PSpice input file shown in Fig. C.1. GNLR is a VCCS connected between nodes 5 and 0, whose controlling nodes are 5 and 0. The equation for the current through GNLR is

$$I = 1*V + 4*V^2 + (-3)*V^3 \text{ amp}$$

App. C: Nonlinear (Polynomial) Dependent Sources

```
NONLINR.CIR    TO MODEL A NONLINEAR RESISTOR
*          A nonlinear VCCS is used to make a
*          voltage-dependent resistor
VSWEEP  5  0  DC  0
*          In the VCCS below (GNLR), the
*          current that flows between nodes
*          8 and 0 IS controlled by the
*          voltage across nodes 8 and 0.
GNLR    5  0  POLY(1)  5  0  0  1  4  -3
.DC VSWEEP  0  1  .01
.PROBE
.END
```

Figure C.1 Input File NONLINR.CIR.

where V is the voltage V(5,0), and I is the current flowing from node 5 to node 0 through GNLR. Figure C.2 illustrates the VCCS and a test circuit to generate an I-V graph.

The graph in Fig. C.3 shows that as the DC voltage source VSWEEP increases from 0 to 1 V, the current I(GNLR) increases from 0 to 2 A. However, the slope is not constant; GNLR behaves as a voltage-dependent, nonlinear resistor.

C.3 NONLINEAR VOLTAGE-CONTROLLED VOLTAGE SOURCES

The general form for a nonlinear VCVS is

EXXXXXX N+ N- POLY(ND) NC1+ NC1- <NC2+ NC2-> . . .
+ P0 P1 . . . <IC = . . . >

where N+ and N- are the nodes to which the voltage source is connected. ND is the number of devices controlling the VCVS,

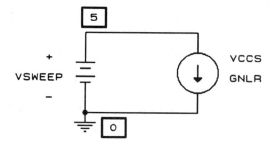

Figure C.2 VCCS Used to Make Nonlinear Resistor.

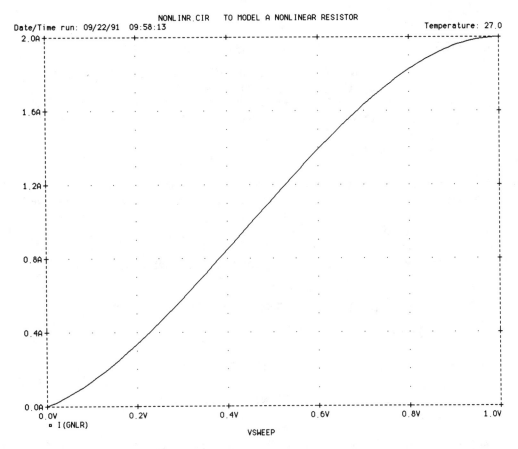

Figure C.3   I-V Curve for Nonlinear Resistor.

and must be specified.   NC+ and NC- are the positive and negative controlling nodes, respectively, and there must be a pair for each dimension.   The initial condition, IC = , is the value of the controlling voltages at time zero.   If initial conditions are not specified the default value is zero.

EXAMPLE:

ESIMPLE 1 4 POLY(1) 7 2 0 0.5 0.3 .06

The voltage across nodes 1 and 4 (produced by VCVS ESIMPLE) is a function of the voltage at node 7 compared to node 2, and has the following equation:

$$V(1,4) = 0.5*V(7,2) + 0.3*V(7,2)^2 + (6E\text{-}2)*V(7,2)^3 \quad \text{volt}$$

App. C: Nonlinear (Polynomial) Dependent Sources   **265**

EXAMPLE:

EPRODUCT 5 7 POLY(2) 20 22 31 33 0 0 0 0 1

     VCVS EPRODUCT produces a voltage between nodes 5 and 7 (positive and negative, respectively) which is the product of the voltages across node pairs 20 & 22 and 31 & 33. Thus the equation for EPRODUCT is

$$V(5,7) = 1.0*V(20,22)*V(31,33) \quad \text{volt}$$

This can be used to make amplitude modulation, either double sideband full carrier (DSB-FC) or double sideband suppressed carrier (DSB-SC). To do this, one of the controlling voltages would be a sinusoid at the carrier frequency, and the other controlling voltage would be the modulating signal plus a constant (for DSC-FC) or without the constant (for DSB-SC).

C.4 NONLINEAR CURRENT-CONTROLLED CURRENT SOURCES

The general form for a nonlinear CCCS is

     FXXXXXX N+ N- POLY(ND) VN1 <VN2 ...>
+ P0 P1 ... <IC = ...>

where N+ and N- are the nodes to which the current source is connected. Positive current is understood to flow from N+ to N- through the CCCS. ND is the number of devices controlling the CCCS, and must be specified. VN1 VN2 ... are the names of the voltage sources through which the controlling currents flow, and there must be one for each dimension. The initial condition, IC = , is the value of the controlling current at time zero. If initial conditions are not specified the default value is zero.

EXAMPLE:

FUNKY 9 3 POLY(1) VINPUT 1 0.4 2

     Current-controlled current source FUNKY is connected between nodes 9 and 3, and is controlled by the current flowing through independent voltage source VINPUT. The equation for the current through FUNKY is

$$I = 1.0 + 0.4*I(VINPUT) + 2*(I(VINPUT))^2 \quad \text{amp}$$

EXAMPLE:

FUZZY 6 8 POLY(3) VA VB VC 0 2 3 4

    CCCS FUZZY causes a current to flow from node 6 to node 8, through FUZZY. The equation for the current through FUZZY is

$$I = 2*VA + 3*VB + 4*VC \quad amp$$

If the P1, P2 and P3 coefficients (2, 3 and 4, respectively) were all made 1, then the current through FUZZY would be the sum of the currents through VA, VB and VC.

C.5 NONLINEAR CURRENT-CONTROLLED VOLTAGE SOURCES

The general form for a nonlinear CCVS is

HXXXXXX N+ N- POLY(ND) VN1 <VN2 . . . >
+ P0 P1 . . . <IC = . . . >

where N+ and N- are the nodes to which the current source is connected. Positive current is understood to flow from N+ to N- through the CCVS. ND is the number of devices controlling the CCVS, and must be specified. VN1 VN2 . . . are the names of the voltage sources through which the controlling currents flow, and there must be one for each dimension. The initial condition, IC = , is the value of the controlling currents at time zero. If initial conditions are not specified the default value is zero.

EXAMPLE:

HITHERE 7 1 POLY(1) VX 0 2 0 4E2

    Current-controlled voltage source HITHERE is connected between nodes 7 and 1, and is controlled by the current through independent voltage source VX. The equation for the voltage across HITHERE is

$$V(7,1) = 0 + 2*I(VX) + 4E2*(I(VX))^3 \quad volt$$

Appendix D

BIBLIOGRAPHY

GENERAL:

Barros, Greg and Joe Domitrowich, "Technology Update: Varied circuit simulators led by Spice," *EDN Product News*, October 1986, pp. 12, 31.

Blume, Wolfram, "Computer Circuit Simulation," *Byte*, July 1986, pp. 165-70.

Chua, Leon O. and Pen-Min Lin, *Computer-aided Analysis of Electronic Circuits: Algorithms & Computational Techniques.* Englewood Cliffs, N.J.: Prentice-Hall, Inc., 1975.

Cuthbert Jr., Thomas R., *Circuit Design Using Personal Computers.* New York, N.Y.: John Wiley & Sons, Inc., 1983.

Epler, Bert, "Circuit Simulation and Modeling Column: SPICE2 Application Notes for Dependent Sources," *IEEE Circuits and Devices Magazine*, Vol. 3, no. 5 (September 1987), pp. 36-44.

268

Giacoletto, L.J., "Simulation and Modeling Column: Alternate PSpice Switch Functions," *IEEE Circuits and Devices Magazine*, Vol. 7, no. 5 (September 1991), pp. 9-37.

Hines, J. Richard, "Selecting a Personal Computer For Circuit Simulation," *VLSI Systems Design*, Vol. VIII, no. 3 (March 1987), pp. 66-71.

Hines, J. Richard, "Reduce simulation times, costs with SPICE convergence aids," *Personal Engineering & Instrumentation News*, May 1987, pp. 47-51.

Nagel, Lawrence W., "SPICE2: A Computer Program to Simulate Semiconductor Circuits," *ERL Memo No. ERL-M520*, Electronics Research Laboratory, University of California, Berkeley, May 1975.

Nurczyk, Mark E., "Analog Circuit Design," *Circuit Cellar Ink,"* Feb./Mar. 1991 (Issue 19), pp. 46-56.

Pederson, Donald O. and Kartikeya Mayaram, *Analog Integrated Circuits for Communication: Principles, Simulation and Design.* Norwell, MA: Kluwer Academic Publishers, 1990.

Radice, Anthony M., "Spice simulations use controlled sources to model NTSC signals," *EDN,* Mar. 1, 1991, pp. 117-130.

Rao, Veerendra and Ronald Hoelzeman, "A SPICE Interactive Graphics Preprocessor," *IEEE Transactions on Education*, Vol. E-29, no. 3 (August 1986), pp. 150-53.

Schreier, Paul G., "Simulators benefit from graphic interfaces, reliable convergence," *Personal Engineering & Instrumentation News*, January 1987, pp. 35-43.

Shear, David, "Board-level analog CAE," *EDN*, May 14, 1987, pp. 138-50.

Sitkowski, Mark, "Simulation and Modeling Column: The Macro Modeling of Phase Locked Loops for the Spice Simulator," *IEEE Circuits and Devices Magazine*, Vol. 7, no. 2 (Mar. 1991), pp. 11-15.

Smith, Craig E. and Ivin L. Holt, "Adding SPICE to the Electronics Industry," *Printed Circuit Design*, Vol. 7, no. 9 (Sept. 1990), pp. 39-47.

Vlach, Jiri and Kishore Singhal, *Computer Methods for Circuit Analysis and Design*. New York, N.Y.: Van Nostrand Reinhold Company, Inc., 1983.

Yang, Ping, "Circuit Simulation and Modeling," *IEEE Circuits and Devices Magazine*, Vol. 3, no. 4 (July 1987), pp. 40-41.

ANALOG BEHAVIORAL MODELING:

Filseth, Eric and Thierry Roullier, "Build Analog Behavioral Models in Six Easy Steps," *Electronic Design*, Vol. 38, no. 22 (Nov. 22, 1990), pp. 105-119.

OPERATIONAL AMPLIFIER MODELING:

Boyle, G.R., B.M. Cohn, D. Pederson and J.E. Solomon, "Macromodeling of Integrated Circuit Operational Amplifiers," *IEEE Journal of Solid-State Circuits*, Vol. SC-9, 1974, pp. 353-64.

"In analog macromodel craze, op amps lead the way," *Electronic Products*, Vol. 32, no. 12 (May 1990), pp. 19-21.

Operational Amplifier Macromodels, Linear Circuits Data Manual, 1990. Dallas, TX.: Texas Instruments Incorporated.

SEMICONDUCTOR MODELING:

Antognetti, P. and G. Massobrio, *Semiconductor Device Modeling with SPICE*. New York: McGraw-Hill Book Company, Inc., 1988.

Bowers, James C. and H.A. Neinhaus, "SPICE2 Computer Models for HEXFETs," Application Note 954A, *HEXFET Power MOSFET Databook, HDB-3*. El Segundo, CA.: International Rectifier Corporation, pp. A-153-60.

Bowers, James C., "Computer design for power electronics," *IEEE Potentials*, May 1986, pp. 36-39.

Early, J.M., "Effects of Space-Charge Layer Widening in Junction Transistors," *Proceedings IRE*, Vol. 46, November 1952, pp. 1401-06.

Ebers, J.J. and J.L. Moll, "Large Signal Behavior of Junction Transistors," *Proceedings IRE*, Vol. 42, December 1954, pp. 1761-72.

Fay, Gary and Judy Sutor, "A Power FET SPICE Model From Data-Sheet Specs," *Powertechnics Magazine*, August 1986, pp. 25-31.

Fay, Gary and Judy Sutor, "Power FET SPICE models are easy and accurate," *Powertechnics Magazine*, Vol. 3, no. 8 (August 1987), 16-21.

Giacoletto, L.J., "Simple SCR and TRIAC PSpice Computer Models," *IEEE Transactions on Industrial Electronics*, Vol. 36, August 1989, pp. 451-455.

Gummel, H.K. and H.C. Poon, "An Integral Charge Control Model for Bipolar Transistors," *Bell System Technical Journal*, Vol. 49, May/June 1970, pp. 827-852.

Herskowitz, Gerald J. and Ronald B. Schilling, *Semiconductor Device Modeling For Computer-Aided Design.* New York: McGraw-Hill Book Company, Inc., 1972.

Macnee, Alan B., "Computer-Aided Optimization of Transistor Model Parameters," *IEEE Transactions on Education*, Vol. E-28, no. 1 (February 1985), 4-11.

Munro, Philip C., "A CAD Model for the BJT with Self Heating," *IEEE Circuits and Devices Magazine*, Vol. 7, no. 3 (May 1991), pp. 7-9.

Schichman, H. and D.A. Hodges, "Modeling and Simulation of Insulated-Gate Field-Effect Transistor Switching Circuits," *IEEE Journal of Solid-State Circuits*, Vol. SC-3, September 1968, pp. 285-89.

Wong, Silphy and Chenmin Hu, "Simulation and Modeling Column: SPICE Macro Model for the Simulation of Zener Diode I-V Characteristics," *IEEE Circuits and Devices Magazine*, Vol. 7, no. 4 (July 1991), pp. 8-52.

SWITCHING POWER SUPPLIES:

Bello, Vincent G., "Computer program adds SPICE to switching-regulator analysis," *Electronic Design*, March 5, 1981, pp. 89-95.

Hageman, Steven, "Behavioral Modeling and PSpice Simulate SMPS Control Loops: Part 1," *Power Conversion Intelligent Motion*, Vol. 16, no. 4, April 1990, pp. 13-24.

TRANSMISSION LINE MODELING:

Tuinenga, Paul W., "Spice simulates lossy transmission," *Electronic Engineering Times*, March 4, 1991.

Appendix E

ANALOG BEHAVIORAL
MODELING

E.1 INTRODUCTION

In addition to allowing users to specify electronic devices in circuits, PSpice has a feature called Analog Behavioral Modeling which allows us to describe a circuit element in terms of its transfer function. Thus, no actual circuit elements (resistors, controlled voltage sources, diodes, etc.) need be present in the input file at all; either a mathematical expression or a look-up table may be used to describe the transfer function of a circuit.

Both nonlinear and linear transfer functions may be described. Nonlinear functions are written as the instantaneous relationship between input and output of a circuit block, while linear functions are described in the frequency domain.

Note: A brief introduction to the PSpice Analog Behavioral Modeling feature is presented here, with examples. For detailed information on this powerful option, refer to the PSpice User's Guide, available from MicroSim Corporation.

E.2 NONLINEAR TRANSFER FUNCTIONS

PSpice uses an extension of the nonlinear (polynomial) dependent sources, which were covered in Appendix C, to make nonlinear circuit blocks which are described by a mathematical relationship between input and output. The voltage-controlled voltage source (VCVS) and the voltage-controlled current source (VCCS) are the only two sources available.

The description can be either a mathematical expression or a table of input-output value pairs. Any such description is thereby independent of frequency.

E.2.1 Nonlinear Transfer Functions By Expression

Nonlinear voltage-controlled voltage sources are described by the following element line:

EXXXXXXN+ N- VALUE = EXPRESSION

where N+ and N- are the nodes to which the voltage source is connected, VALUE is the voltage produced by EXXXXXX, with the N+ node positive compared to the N- node, and EXPRESSION is a mathematical relationship which may involve both currents and voltages.

EXAMPLE:

E-AM 3 0 Value = {(1+1*sin(9K*time))*(10*sin(250K*time))}

The voltage source connected between nodes 3 and 0 is controlled by the product of a 1 Vp sinusoid at 8K rad/sec with an offset voltage of 1 VDC, and a 10 Vp sinusoid at 250K rad/sec. This produces a 100% modulated AM carrier, similar to the one presented in Section 12.5.9. The braces enclosing the expression must appear, and note the use of the variable "time" in the expression.

A PSpice input file that will graph the AM carrier described above is shown in Fig. E.1.

```
AM-EASY.CIR  Amplitude Modulation Done With Analog Behavioral Modeling
E-AM  30  0  Value = {(1+1*sin(9K*time))*(10*sin(250K*time))}
RDUMMY  30  0  1
.TRAN  2U  4M  0  2U
.OPTIONS ITL5=0
.PROBE
.END
```

Figure E.1 Input File AMEASY.CIR.

Nonlinear voltage-controlled current sources are described by the following element line:

GXXXXXXN+ N- VALUE = EXPRESSION

where N+ and N- are the nodes to which the current source is connected, VALUE is the current flowing from the N+ node to the N- node, through the VCCS, and EXPRESSION is a mathematical relationship which may involve both currents and voltages. The EXPRESSION has to fit on a single line.

EXAMPLE:

G-PWR-DBm 11 23 Value = {30 + 10*LOG(V(1,2)*I(RL))}

The current source connected between nodes 11 and 23 is a current equal to the power in dBm produced by the product of the voltage V(1,2) and the current through resistor RL. The constant 30 in the expression converts dBW to dBm.

E.2.2 Nonlinear Transfer Functions By Table

A table of input-output pairs can also be used to describe nonlinear transfer functions. The general format for such a VCVS or VCCS is:

EXXXXXX N+ N- TABLE {EXPRESSION} = (IN1,OUT1)
+ (IN2,OUT2) (IN3,OUT3) . . .

GXXXXXX N+ N- TABLE {EXPRESSION} = (IN1,OUT1)
+ (IN2,OUT2) (IN3,OUT3) . . .

Applications of a nonlinear transfer function based on data in a table include using data obtained by laboratory measurement, or simulating a desired input-output relationship (such as an ideal diode).

EXAMPLE:

G-diode 6 17 table {V(6,17)} = (0V,0A) (.59V,0A) (.6V,4A)

The current source connected between nodes 6 and 17 (with positive current assumed to flow from 6 to 17) is determined by the table of three input-output value pairs. A semi-ideal diode is described, in that there is an offset voltage of 0.6 V.

An important characteristic of transfer functions described by tables is that for any value of input variable less than the lowest defined, the output variable is a constant (the output value in the pair containing the lowest defined input variable). Conversely, for any input variable greater than the largest defined, the output variable is a constant (the output value in the pair containing the largest defined output variable). Thus, in the example above if the voltage V(6,17) is less than 0 V, the current is constant at 0 A, and if V(6,17) is greater than .6 V, the current is fixed at 4 A.

In a graph of the transfer function, Fig. E.2, we can see that as V(6,17) rises from 0 V to 590 mV, the current is 0 A. From 590 mV to 600 mV, the current rises steeply with voltage, and for voltages above 600 mV, the current is fixed at 4 A.

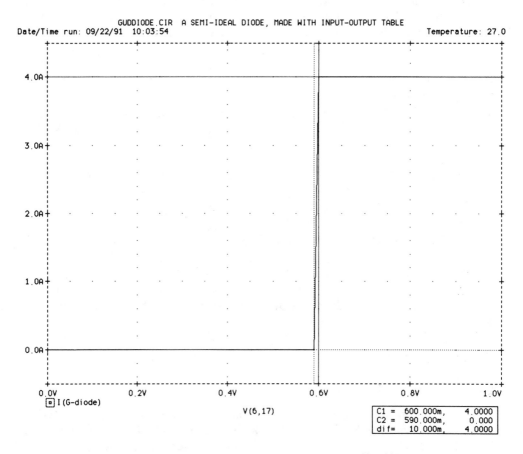

Figure E.2 I-V Relationship Described by a Nonlinear Transfer Function Table.

Analog Behavioral Modeling   App. E:

E.3 LINEAR TRANSFER FUNCTIONS

The PSpice linear device models using Analog Behavioral Modeling, as with the nonlinear transfer functions, use VCVS and VCCS sources. Linear transfer functions may be expressed either as a mathematical expression which involves a Laplace transform, or as a table of frequency response data. Unlike nonlinear transfer functions, linear transfer functions do not specify an instantaneous relationship between input and output. The output depends not only on the instantaneous input value, but also on what has happened to the circuit prior to the input being applied.

E.3.1 Linear Transfer Function By Expression

Linear voltage-controlled sources are described by the following element lines:

    Exxxxxx N+ N- Laplace {EXPRESSION} =
    + {Transform}

    Gxxxxxx N+ N- Laplace {EXPRESSION} =
    + {Transform}

where N+ and N- are the nodes to which the controlled voltage (E) or current (G) source is connected. The word Laplace must appear followed by a space, EXPRESSION (in braces) is the input and may consist of any combination of voltages and currents, and the output voltage (E) or current (G) is determined by the Transform (in braces) of the input. EXPRESSION and Transform each must fit on a single line in the input file.

EXAMPLE:

E-RLC 20 0 Laplace {V(8)} =
+ {1/(2.5E-12*s*s + 2E-7*S + 1)}

The value of the voltage source connected between nodes 20 and 0 is determined by the second-order Laplace transform of the voltage at node 8.

In order to illustrate how this works in a PSpice analysis, refer to the circuit diagram in Fig. E.3. An AC voltage source, VIN, is connected to node 8 with its dummy resistor RIN. The node 8 voltage is the input to E-RLC, a VCVS which has the transfer function above. Figure E.4 is the PSpice in-

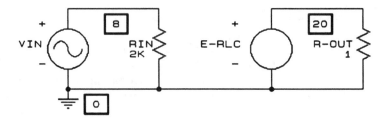

Figure E.3 Circuit for Laplace Transform Voltage-Controlled Voltage Source.

```
LAPLACE.CIR  Linear Transfer Function, Using Laplace Expression
VIN  8  0  AC  1
RIN  8  0  2K
E-RLC  20  0  Laplace {V(8)} =
+ {1/(2.5E-12*S*S + 2E-7*S + 1)}
R-OUT  20  0  1
.AC DEC 100  10K  1MEG
.PROBE
.END
```

Figure E.4 Input File LAPLACE.CIR.

put file LAPLACE.CIR, which sweeps VIN from 10 KHz to 1 MHz. The magnitude and phase of the output at node 20 are graphed in Fig. E.6. Note from both the graphs that this is the response of a circuit which is resonant at 100 KHz.

The same circuit was used in the input file LAPLACE2.CIR, in which a 3 microsecond voltage pulse replaces the AC voltage. It is described in Fig. E.6, in which a transient analysis is specified. The input pulse and the output voltage, a decaying sinusoid, are seen in Fig. E.7. Clearly an underdamped second order system has been described by the linear transfer function E-RLC.

E.3.2 Linear Transfer Function By Frequency Response Table

Linear voltage-controlled sources are described by the following element lines:

Exxxxxx   N+ N- FREQ {EXPRESSION} =
+ (Freq1 MagDB1 Phase1) (Freq2 MagDB2 Phase2) . . .

Gxxxxxx   N+ N- FREQ {EXPRESSION} =
+ (Freq1 MagDB1 Phase1) (Freq2 MagDB2 Phase2) . . .

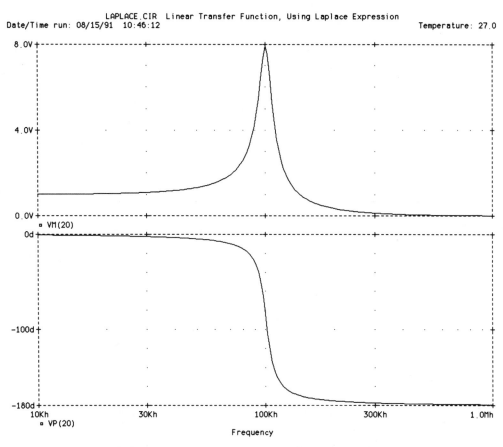

Figure E.5 Laplace Circuit Output Voltage Magnitude and Phase vs. Frequency.

```
LAPLACE2.CIR  Linear Transfer Function, Using Laplace Expression
VIN  8  0  PULSE (0  1  2U  1N  1N  3U)
RIN  8  0  2K
E-RLC  20  0  Laplace {V(8)} =
+ {1/(2.5E-12*S*S + 2E-7*S + 1)}
R-OUT  20  0  1
.TRAN  .5U  80U  0  .2U
.PROBE
.END
```

Figure E.6 Input File LAPLACE2.CIR.

where N+ and N- are the nodes to which the controlled voltage (E) or current (G) source is connected. The word FREQ must appear followed by a space, EXPRESSION (in braces) is the input and may consist of any combination of voltages and currents, and the output voltage (E) or current (G) is determined by the frequency response data which follows. EXPRESSION must fit on

App. E: Analog Behavioral Modeling   279

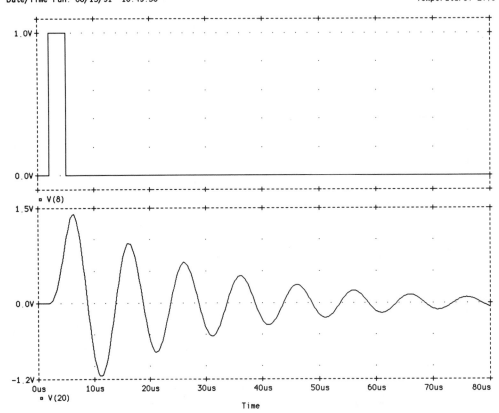

**Figure E.7 Input Current Pulse and Output
Voltage for Laplace Circuit.**

```
FREQTAB.CIR  Linear Transfer Function, Using Frequency Response Table
IIN  0  7  AC  1
VDUMMY  7  0  DC  0
R-FILTER  20  0  1
E-FILTER  20  0  FREQ {I(VDUMMY)} =
+ (1.000E+01  -7.161E+01   1.798E+02) (1.778E+01  -6.161E+01   1.796E+02)
+ (3.162E+01  -5.160E+01   1.793E+02) (5.623E+01  -4.158E+01   1.787E+02)
+ (1.000E+02  -3.153E+01   1.777E+02) (1.778E+02  -2.135E+01   1.758E+02)
+ (3.162E+02  -1.077E+01   1.719E+02) (5.623E+02   1.281E+00   1.616E+02)
+ (1.000E+03   1.630E+01   8.848E+01) (1.778E+03   1.112E+01   1.804E+01)
+ (3.162E+03   9.120E+00   7.951E+00) (5.623E+03   8.552E+00   4.178E+00)
+ (1.000E+04   8.379E+00   2.302E+00)
.AC DEC 20 10 100K
.PROBE
.END
```

Figure E.8 Input File FREQTAB.CIR.

a single line in the input file, the magnitude data in the table (e.g. MagDB1) must be in decibels, and the frequencies in the table must be in ascending order.

EXAMPLE:

Exxxxxx N+ N- FREQ {EXPRESSION} =
+ (Freq1 MagDB1 Phase1) (Freq2 MagDB2 Phase2) . . .

E-RES 26 14 FREQ {V(9)} =
+ (500 .0873 -.009) (5K 35.96 -90)
+ (50K -.399 -179.9)

VCVS E-RES is connected to nodes 26 and 14. Its input is the voltage at node 9, and its output is determined by the frequency table data for three frequencies. For example, with a 1 V AC source connected to node 9, at 5 KHz the voltage between nodes 26 and 14 is 35.96 dB above 1 V with a phase angle of -90 degrees.

Figure E.8 is the PSpice input file FREQTAB.CIR which uses a frequency table with magnitude and phase data at 13 frequen-

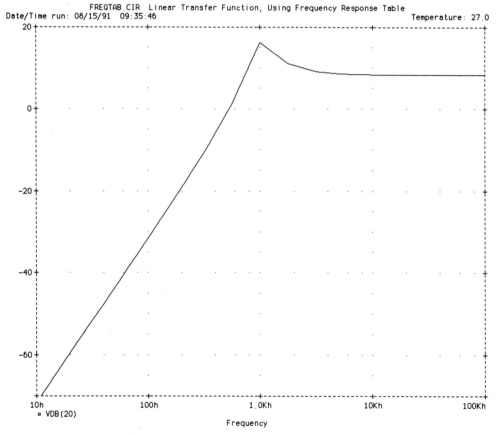

**Figure E.9 Bode Plot of Filter Response
Described by Frequency Table.**

```
FREQTAB2.CIR  Linear Transfer Function, Using Frequency Response Table
VIN  10  0  PULSE (0  1  .5M  1U  1U)
RIN  10  0  1K
R-FILTER  20  0  1
E-FILTER  20  0  FREQ {V(10)} =
+ (1.000E+01  -7.161E+01   1.798E+02) (1.778E+01  -6.161E+01   1.796E+02)
+ (3.162E+01  -5.160E+01   1.793E+02) (5.623E+01  -4.158E+01   1.787E+02)
+ (1.000E+02  -3.153E+01   1.777E+02) (1.778E+02  -2.135E+01   1.758E+02)
+ (3.162E+02  -1.077E+01   1.719E+02) (5.623E+02   1.281E+00   1.616E+02)
+ (1.000E+03   1.630E+01   8.848E+01) (1.778E+03   1.112E+01   1.804E+01)
+ (3.162E+03   9.120E+00   7.951E+00) (5.623E+03   8.552E+00   4.178E+00)
+ (1.000E+04   8.379E+00   2.302E+00)
.TRAN  40U  4M  0  20U
.PROBE
.END
```

Figure E.10 Input File FREQTAB2.CIR.

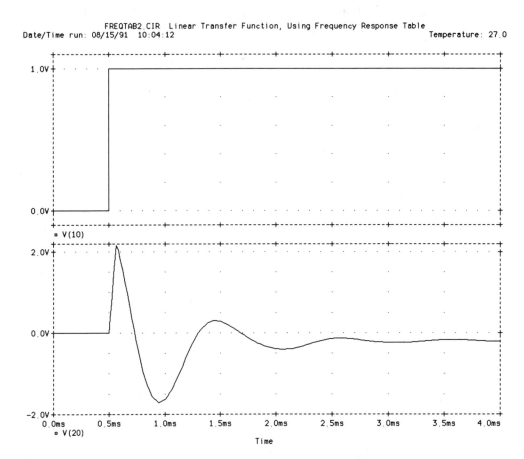

**Figure E.11 Input and Output Voltages
From Filter Described by Frequency Table.**

cies. The data represents the frequency response of a second-order active high-pass filter with a cutoff frequency of 1 KHz. The input to VCVS E-FILTER is the current through dead voltage source VDUMMY, which is supplied by AC current source IIN. Figure E.9 shows the Bode plot of the filter response, with a peak in the response at 1 KHz, indicating that the high-pass filter has a Chebyshev response.

The same frequency table response is subjected to a voltage step in the input file FREQTAB2.CIR, Fig. E.10. To illustrate the difference in possible types of inputs, here the input to the VCVS is voltage source VIN, a voltage step. A transient analysis is specified, and as the results graphed in Fig. E.11 show, the high-pass filter responds to the voltage step (top plot) by ringing. The damped sinusoid in the lower plot has a period of 1 millisecond, corresponding to the 1 KHz cutoff frequency of the filter.

Appendix F

QUICK REFERENCE TABLES - ELEMENT LINES AND CONTROL LINES

F.1 QUICK REFERENCE TABLES

Two reference tables follow: one for element lines (resistors, capacitors, etc.) and one for control lines (.AC, .PROBE, .TF, etc.). Each table entry contains condensed information on how to use the lines most often needed by users of PSpice. For complete information on that line, refer to the appropriate chapter. Advanced users should consult the PSpice User's Guide, available from MicroSim Corporation.

Parameters shown below in < > are optional.

F.2 Element Lines Reference Table

B **GaAsFET** Bxxxxxx **ND NG NS MODNAME <AREA>**
    Example: B_RFAMP 6 2 4 HOTSTUFF 1.7
                 .MODEL HOTSTUFF GASFET(level=2 vto=-3)

C **Capacitor** Cxxxxxx **N+ N- VALUE <IC = INCOND>**
    Example: CFILTER 14 22 33U IC=12V

D	Diode	Dxxxxxx N+ N- modname <area> <off> <IC=VD>
	Example:	Drectify 7 15 hipower
		.MODEL hipower D(RS=.5 BV=400 IBV=50M)

E	VCVS	Exxxxxx N+ N- NC+ NC- VALUE
	Example:	EGAIN 14 9 6 2 250

E	VCVS	Exxxxxx N+ N- poly(nd) NC1+ NC1- . . . coeff
	Example:	Eproduct 5 7 poly(2) 20 22 31 33 0 0 0 0 1

E	VCVS	Exxxxxx N+ N- VALUE = {expression}
	Example:	Eam 3 0 value={sin(3*time)*(1+cos(9*time))}

E	VCVS	Exxxxxx N+ N- TABLE {expression}=(in1,out1). . .
	Example:	E2 7 4 table {V(5)}=(0,0) (.6,0) (.61,5)

E	VCVS	Exxxxxx N+ N- LAPLACE {expression}=
		+ {transform}
	Example:	E-RLC 20 0 Laplace {V(8)} =
		+ {1/(2.5E-12*s*s + 2E-7*S + 1)}

E	VCVS	Exxxxxx N+ N- FREQ {expression}=
		+ (freq1 magdb1 phase1)
		+ (freq2 magdb2 phase2) . . .
	Example:	E-RES 26 14 FREQ {V(9)} =
		+ (500 .0873 -.009)(5K 35.96 -90)
		+ (50K -.399 -179.9)

F	CCCS	Fxxxxxx N+ N- VNAME VALUE
	Example:	FOUT 37 24 VSENSE 5

F	CCCS	Fxxxxxx N+ N- poly(nd) vnam1 vnam2 . . . coeff
	Example:	Fuzzy 6 8 poly(3) VA VB VC 0 2 3 4

G	VCCS	Gxxxxxx N+ N- NC+ NC- VALUE
	Example:	GGAIN 14 9 6 2 250

G	VCCS	Gxxxxxx N+ N- poly(nd) NC1+ NC1- . . . coeff
	Example:	Gproduct 5 7 poly(2) 20 22 31 33 0 0 0 0 1

G	VCCS	Gxxxxxx N+ N- VALUE = {expression}
	Example:	Gam 3 0 value={sin(3*time)*(1+cos(9*time))}

G	VCCS	Gxxxxxx N+ N- TABLE {expression}=(in1,out1). . .
	Example:	G2 7 4 table {V(5)}=(0,0) (.6,0) (.61,5)

G   VCCS   Gxxxxxx N+ N- LAPLACE {expression}=
                 + {transform}
     Example: G-RLC 20 0 Laplace {V(8)} =
                 + {1/(2.5E-12*s*s + 2E-7*S + 1)}

G   VCCS   Gxxxxxx N+ N- FREQ {expression}=
                 + (freq1 magdb1 phase1)
                 + (freq2 magdb2 phase2) . . .
     Example: G-RES 26 14 FREQ {V(9)} =
                 + (500 .0873 -.009)(5K 35.96 -90)
                 + (50K -.399 -179.9)

H   CCVS   Hxxxxxx N+ N- VNAME VALUE
     Example: HOUT 37 24 VSENSE 5

H   CCVS   Hxxxxxx N+ N- poly(nd) vnam1 vnam2 . . . coeff
     Example: Happy 6 8 poly(3) VA VB VC 0 2 3 4

I   Current Source   Ixxxxxx N+ N- DC/tran-value
                            Ixxxxxx N+ N- AC ACMag ACphase
                            Ixxxxxx N+ N- Trankind (tpar1 tpar2 . . .)
     Examples: I_owe 4 5 DC +3
                 Ibase 0 7 AC 40E-3 60
                 I_HZ 2 9 SIN(20M 80M 10KHz)
     Trankind is one of the following:
     PULSE(V1 V2 timedelay risetime falltime pulsewidth period)
     SIN(VO VA freq timedelay dampfactor phase)
     EXP(V1 V2 timedelay1 riseTC1 timedelay2 riseTC2)
     PWL(T1 V1 T2 V2 T3 V3 . . .)
     SFFM(VO VA carrier-freq mod-index signal-freq)
     *A single current source can have a DC value, an AC value
     and a Trankind simultaneously.*

J   JFET Jxxxxxx ND NG NS modname <area> <off> <ic=vds,vgs>
     Example: Jbuffer 8 5 7 Easy
                 .MODEL easy PJF

K   Mutual Inductance   Kxxxxxx Lyyyyyy Lzzzzzz Value
     Example: KPRISEC LPRI LSEC 0.97

L   Inductor Lxxxxxx N+ N- Value <IC=Incond>
     Example: LTANK 11 34 560U 2mA

M   MOSFET Mxxxxxx ND NG NS NB modname <optional params.>
     Example: Mbuffer 8 5 7 0 Simple
                 .MODEL simple NMOS

Q BJT Qxxxxxx NC NB NE <NS> modname <area> <off>
+ <IC=vbe,vce>
Example: Qoutput 8 5 7 0 Plain
.MODEL Plain NPN(BF=80)

R Resistor Rxxxxxx N1 N2 Value <TC=TC1<,TC2>>
Example: Realbig 74 26 100E9
Note: resistors can also have a Model Name between N2 and Value above; if so, a .MODEL line for the resistor model must be present. See Appendix A.7 and A.8.

S Voltage-controlled Switch
Sxxxxxx N1 N2 NC+ NC- Modelname
Example: S/fuse 21 11 7 0 sloblo
.MODEL sloblo vswitch ron=1 roff=1G von=4 voff=3

T Transmission Line
TXXXXXX NA+ NA- NB+ NB- Z0=ZVAL F=FREQ
+ <NL=NLENGTH> <IC = VA, IA, VB, IB>
or
TYYYYYY NA+ NA- NB+ NB- Z0=ZVAL TD=TVALUE
+ <IC = VA, IA, VB, IB>
Examples: TSTUB 1 2 3 4 Z0 = 50 TD = 60N
or
TSTUB 1 2 3 4 Z0 = 50 F = 30MEG NL = 1.8

V Voltage Source Vxxxxxx N+ N- DC/tran-value
Vxxxxxx N+ N- AC ACMag ACphase
Vxxxxxx N+ N- Trankind (tpar1 tpar2 . . .)
Examples: V_owe 4 5 DC +3
Vbase 0 7 AC 40E-3 60
V_HZ 2 9 SIN(20M 80M 10KHz)
Trankind is one of the following:
PULSE(V1 V2 timedelay risetime falltime pulsewidth period)
SIN(VO VA freq timedelay dampfactor phase)
EXP(V1 V2 timedelay1 riseTC1 timedelay2 riseTC2)
PWL(T1 V1 T2 V2 T3 V3 . . .)
SFFM(VO VA carrier-freq mod-index signal-freq)
A single voltage source can have a DC value, an AC value and a Trankind simultaneously.

W Current-controlled Switch
Wxxxxxx N1 N2 Vname Modelname
Example: W/fuse 21 11 Vsensing fasblo
.MODEL fasblo iswitch ron=1 roff=1G von=4 voff=3

App. F: Quick Reference Tables   287

X	Subcircuit Call	Xyyyyyy N1 <N2 N3 . . . > SUBNAME
	Example:	X3 10 20 RCFILTER

F.3   Control Lines Reference Table

.AC	AC Analysis	.AC LIN NP FSTART FSTOP
		.AC DEC ND FSTART FSTOP
		.AC OCT NO FSTART FSTOP

Example: .AC LIN 101 5000 6000

.DC   DC Analysis
.DC SRC START STOP INCR <SRC2 START2 STOP2 INCR2>
.DC DEC SRC START STOP ND <SRC2 START2 STOP2 ND>
.DC OCT SRC START STOP NO <SRC2 START2 STOP2 NO>
.DC SRC LIST VALUE1 VALUE2 . . .
Note: For detailed information on alternate formats for DC analysis control lines, refer to the PSpice User's Guide, available from MicroSim Corporation.
Examples: .DC VGATE -2 3 0.5
                   .DC IBASE 0 100U 10U VCE 0 20 4

.END   Marks the end of a circuit in a PSpice input file.

.ENDS   Marks the end of a subcircuit in a Pspice input file.
Example: .ENDS RCFILTER

.FOUR	Fourier Analysis	.FOUR FREQ OV1 <OV2 OV3 . . .>
	Example:	.FOUR 60 V(6) I(VSOURCE)

.IC   Initial Bias Conditions
                .IC V(NODENUM1) = VAL1 V(NODENUM2) = VAL2 . . .
Example: .IC V(12) = -5 V(8) = 12

.LIB	Library File	.LIB "FILENAME.LIB
	Example:	.LIB "EVAL.LIB"

.MODEL Model
     .MODEL MODNAME TYPE<(PAR1=PVAL1 PAR2=PVAL2 . . .)>
     Examples: R1 7 8 DELTA 100K
                      .MODEL DELTA RES(TC1 = -.003)

                      COUT 4 6 DRIFTY 10U
                      .MODEL DRIFTY CAP(VC1 = .02)

.NODESET Sets initial value(s) of node voltages.
       .NODESET V(NODENUM1) = VAL1 V(NODENUM2) = VAL2 . . .
    Example: .NODESET V(12) = -5 V(8) = 12

.NOISE Noise Analysis   **.NOISE OUTPUTV INPUTSRC NUMSUM**
    Example: .NOISE V(2,3) VIN1 10

.OP   Operating Point   **.OP**
    Example: .OP

.OPTIONS   Options
       .OPTIONS OPT1 OPT2 . . . (or OPT = OPTVAL . . .)
    Examples: .OPTIONS NOPAGE NODE LIMPTS = 501
               .OPTIONS TNOM = 0 ACCT RELTOL = .01 ITL5 = 0

.PLOT   **Graph Analysis Results**
    .PLOT ANALTYPE OV1 <(PLO1,PHI1)> <OV2 <(PLO2,PHI2)>
    + . . . OV8 <(PLO8,PHI8)>>
    Examples: .PLOT DC V(1) (-4,12) V(5,7) I(VGATE)
              .PLOT TRAN V(6) I(VSENSING) V(4,2) (0,8)
              .PLOT AC VM(3) VP(3) (10,50) VG(6,4)
              .PLOT NOISE INOISE ONOISE(DB)

.PROBE   **Write Analysis Results to file PROBE.DAT**
    .PROBE <OV1> <OV2> . . .
    Examples: .PROBE
              .PROBE IP(VSOURCE) V(16)

.SENS   Sensitivity Analysis   **.SENS OV1 <OV2> . . .**
    Example: .SENS V(5) I(VSUPPLY)

.STEP   **Change Values in Parametric Analysis**
    .STEP TYPE MODNAME(Prefix) Startval Stopval Incrval
    Examples: COUT 4 6 DRIFTY 10U
                  .MODEL DRIFTY CAP
                  .STEP CAP DRIFTY(C) .6 1.5 .3

                  RIN 70 75 RPOT 1
                  .MODEL RPOT RES
                  .STEP DEC RES RPOT(R) 10 1MEG 1

.SUBCKT   **Defines Subcircuit**
    .SUBCKT SUBNAME N1 <N2 N3 . . . >

```
Example:   .SUBCKT  TEE  1  2
           RLEFT  1  8  60
           RRIGHT  8  2  80
           RMID    8  0  50
           CRIGHT  2  0  1U
           .ENDS   TEE
```

.TEMP   **Temperature**   **.TEMP T1 <T2 <T3 . . . >>**
     Example: .TEMP 0 25 60 100

.TF   **Transfer Function Analysis**
          .TF OUTPUTVAR INPUTSRC
     Example: .TF V(5) VIN

.TRAN   **Transient Analysis**
          .TRAN TSTEP TSTOP <TSTART <TMAX>> <UIC>
     Examples: .TRAN 1U 80U
               .TRAN 5M 500M 100M 2M UIC

.WIDTH   **Sets Width of PSpice Output File**
          .WIDTH OUT = COLNUMOUT
     Example: .WIDTH OUT = 132 *(default width is 80)*

*   **Comment Line   ***
     Example: *Subcircuit below uses Boyle op-amp model

;   **End of Line Comment Line   ;**
     Example: L-tank 3 4 0.05 ; this has a ferrite core

Appendix G

OPERATIONAL AMPLIFIERS & PSpice

G.1 INTRODUCTION

The operational amplifier (op-amp) is a basic building block used in electronic circuitry for many purposes, including amplifiers, comparators, active filters, oscillators, integrators, differentiators, summing amplifiers, signal conditioners, and digital to analog and analog to digital converters. In order to simulate the operation of a circuit which uses one or more op-amps, a model for the op-amp must be found which is adequate for the conditions under which the op-amp will be used.

For example, if the op-amp will be used in a cardiac pacemaker inside a human chest, the effect of thermal drift on circuit performance is of little concern. It is unnecessary, and indeed wasteful of time for both model development and circuit analysis, to include thermal effects in an op-amp model for a pacemaker. In contrast, an op-amp used in the engine compartment of a car must function over an extremely wide range of ambient temperatures, and thermal effects must be a part of an accurate op-amp model used in automotive applications.

Similarly, when an op-amp is being used as a DC amplifier, monitoring a liquid level or the temperature of a large mass like a human body or an engine block, the frequency response and slew rate of the op-amp are of little consequence. The point is, the op-amp model used for circuit simulation has to be good enough to predict accurately how the circuit will behave. In the following pages, four models for op-amps will be presented with a discussion of the limitations of each.

Modeling an op-amp using all of its circuit elements is not only a daunting task but also can result in analysis times that are excessive. In fact, it is all too easy to run out of memory on your computer if you use op-amp models which contain more circuit elements than is necessary. A macromodel is a circuit model which uses far fewer circuit elements than the op-amp itself while still allowing PSpice to predict accurately (we hope!) the behavior of an op-amp circuit. In this appendix, progressively more complex macromodels are presented, and sources of op-amp macromodels are given.

Of course, it is vital to build and thoroughly test a prototype of any op-amp circuit after computer simulation has been completed to confirm both the soundness of the circuit design and the adequacy of the op-amp macromodel.

G.2 BARE BONES OP-AMP MODEL

When an op-amp circuit is operated in the linear region (where it never saturates), and slew rate and gain-bandwidth effects are of little importance, the model shown in Fig. G.1 may be appropriate. Typically, the input resistance, RIN, is 1E6 ohms or more, and AOL (the open-loop voltage gain) is at least 20E3 V/V. The output of the op-amp, a voltage-controlled voltage source, is the open-loop gain multiplied by the differential input voltage, VID. While this is a very simple model which neglects almost every limitation of a real op-amp, it

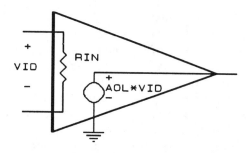

Figure G.1 Bare Bones Op-Amp Model.

will allow PSpice to predict accurately the behavior of many circuits in which the op-amp's job is to provide a big chunk of voltage gain with a large input impedance. See Examples 12.4.7 and 12.5.8 where this model is used.

G.3 OP-AMP MODEL WITH FREQUENCY EFFECTS

Figure G.2 is an improvement over the bare bones model in that a single-pole lowpass response is provided, and an output impedance is added. The input impedance now includes a 2E6 ohm differential input resistance, and a 250E6 ohm common-mode input impedance (created by the two 500E6 ohm resistors). Two voltage-controlled current sources provide the DC voltage gain: [(2 A/A)*1K ohm]*[(1.33 A/A)*75 ohm] = 200E3 V/V. The gain-bandwidth is determined by the 1K ohm resistor and the 26.6 uF capacitor, which have a breakpoint of 6 Hz, typical of the open-loop bandwidth of a general purpose op-amp like a 741.

Missing from this model is saturation; the op-amp output voltage swing is limited only by the imagination of the user!

G.4 BOYLE OP-AMP MODEL

A paper written by Boyle, Cohn, Pederson and Solomon in the IEEE Journal of Solid State Physics in 1974 presented a macro-model for op-amps which was a significant improvement over previous models. (See Appendix D.) It has become known as the Boyle model, and includes frequency effects, DC effects and the ability of the output stage of the op-amp to saturate. A model close to the original Boyle model is shown in Fig. G.3,

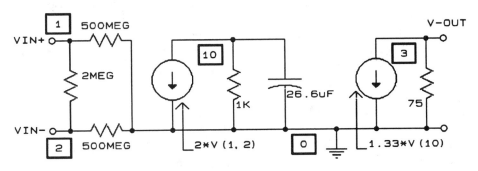

Figure G.2 Op-Amp Model Including Single-Pole Frequency Effects.

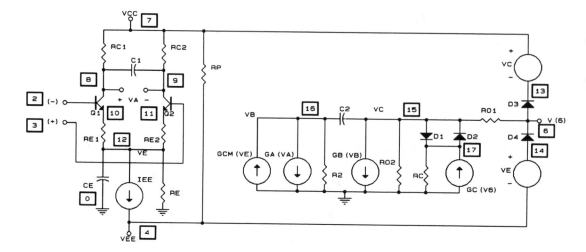

Figure G.3 Boyle Model for Op-Amp.

in which we can see that actual bipolar junction transistors are used as the input differential amplifier. This model thereby provides input bias currents (the base currents of Q1 and Q2).

The op-amp output (node 6) is tied with clamping diodes D3 and D4 to the positive and negative supply voltages through voltage sources VC and VE respectively. VC and VE are typically 2 or 3 volts, which means that the peak output voltage swing in either direction is limited to the supply voltage less 2 or 3 volts. Unfortunately, when the op-amp is driven into saturation, D3 and D4 start conducting and some rather huge currents can "flow" into the VCC and VEE power supplies as a result. While these currents exist only in the memory of the computer, this way of limiting the peak voltage swing provides unrealistically high supply currents when the op-amp is saturated. Three capacitors in the model provide the frequency effects; C1 and C2 provide lowpass breakpoints for the differential gain, while CE causes the common-mode gain to increase with frequency.

G.5   PSpice OP-AMP MODEL

An extension of the Boyle op-amp model is shown in Fig. G.4. Like the Boyle model, it is a macromodel which uses two transistors at the input and a collection of controlled sources to simulate the operation of the actual op-amp. It does offer a

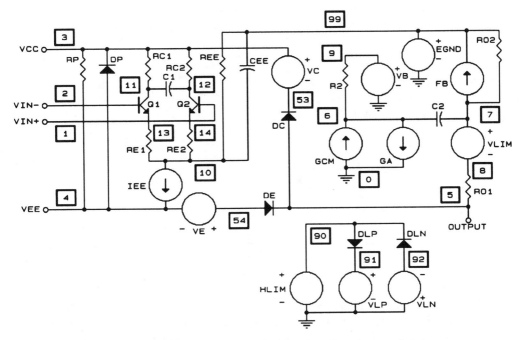

Figure G.4 PSpice Model for Op-Amp with NPN Input Transistors.

significant improvement over the Boyle model in that when the output saturates, large supply currents do not flow. The PSpice op-amp macromodel, when connected in a circuit as a subcircuit, will have supply currents, input bias currents, input impedance, differential- and common-mode gain, open loop gain (magnitude and phase) vs. frequency, and output voltage and current limiting typical of the actual op-amp. The advantage of using the macromodel is that vastly less computer memory and computing time are needed when simulating op-amp circuits.

Figure G.4 is the PSpice macromodel for an op-amp with NPN transistors at the input. Information on models using PNP or FET input transistors can be found in the PSpice User's Guide and from the sources listed in Section G.6.

G.5.1 Closed-Loop Gain Using PSpice Op-Amp Model

The PSpice evaluation version comes with a reduced library of components, including two op-amps: the LM324 and the uA741. Figure G.5 is a test circuit, using the PSpice uA741 op-amp macromodel, in which the input to the inverting amplifier cir-

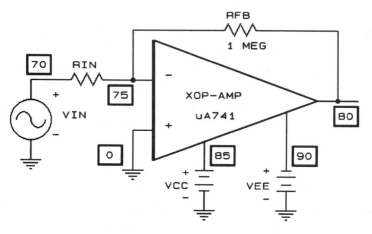

Figure G.5 PSpice uA741 Test Circuit.

cuit is an AC voltage source and the input resistor, RIN, will be varied to change the closed-loop gain. The PSpice input file 741-1.CIR (Fig. G.6) shows the subcircuit for the uA741 between the two dashed lines. RIN is defined by three lines which make RIN vary from 10 ohms to 1 megohm, in decade steps. This will make the closed-loop gain vary from 1E5 V/V (when RIN = 10 ohms) to 1.0 V/V (when RIN = 1 megohm). AC voltage source VIN is varied in frequency from 1 Hz to 1 MHz, with 5 steps per decade. The Bode plot in Fig. G.7 has a family of curves, each representing the gain vs. frequency for a particular RIN value and closed-loop gain.

G.5.2 Saturation Using PSpice Op-Amp Model

The circuit of Fig. G.5 can also be used to show the effect of driving the op-amp output into saturation. Figure G.8 is the input file 741-2.CIR, in which resistor RIN is stepped from 0.4 of its nominal value of 100K ohms to 1.0 of its nominal value. In all, RIN is set to 40K, 60K, 80K and 100K ohms, which makes the inverting gain magnitude equal to 25, 16.7, 12.5 and 10 V/V respectively. VIN is a 1 Vp, 10 Hz sinusoid, and the output voltage, V(80), is graphed in Fig. G.9. The output is sinusoidal when the gain is 10 V/V and 12.5 V/V, but saturation is clearly evident in the clipping of the output sinusoid with gains of 16.7 and 25 V/V.

```
741-1.CIR  use of PSpice Model for uA741 Op-Amp
*        will make a family of freq. response curves
*        for varying closed-loop gains
VIN  70  0  AC  1
RIN  70  75  RPOT 1
.MODEL RPOT RES
.STEP DEC RES RPOT(R) 10 1MEG 1
RFB  80  75  1MEG
VCC  85  0   15
VEE  90  0   -15
XOP-AMP  0  75  85  90  80  UA741
*------------------------------------------------------------------
* connections:   non-inverting input
*                 | inverting input
*                 | | positive power supply
*                 | | | negative power supply
*                 | | | | output
*                 | | | | |
.subckt uA741    1 2 3 4 5
  c1   11 12 8.661E-12
  c2    6  7 30.00E-12
  dc    5 53 dx
  de   54  5 dx
  dlp  90 91 dx
  dln  92 90 dx
  dp    4  3 dx
  egnd 99  0 poly(2) (3,0) (4,0) 0 .5 .5
  fb    7 99 poly(5) vb vc ve vlp vln 0 10.61E6 -10E6 10E6 10E6 -10E6
  ga    6  0 11 12 188.5E-6
  gcm   0  6 10 99 5.961E-9
  iee  10  4 dc 15.16E-6
  hlim 90  0 vlim 1K
  q1   11  2 13 qx
  q2   12  1 14 qx
  r2    6  9 100.0E3
  rc1   3 11 5.305E3
  rc2   3 12 5.305E3
  re1  13 10 1.836E3
  re2  14 10 1.836E3
  ree  10 99 13.19E6
  ro1   8  5 50
  ro2   7 99 100
  rp    3  4 18.16E3
  vb    9  0 dc 0
  vc    3 53 dc 1
  ve   54  4 dc 1
  vlim  7  8 dc 0
  vlp  91  0 dc 40
  vln   0 92 dc 40
.model dx D(Is=800.0E-18 Rs=1)
.model qx NPN(Is=800.0E-18 Bf=93.75)
.ends
*------------------------------------------------------------------
.AC DEC 5 1 1MEG
.PROBE
.END
```

Figure G.6 Input File 741-1.CIR.

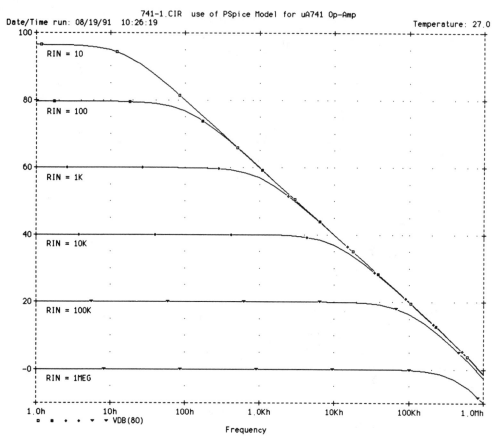

**Figure G.7 Op-Amp Circuit Frequency
Response with Varying Closed-Loop Gains.**

G.5.3   Slew-Rate Limiting With PSpice Op-Amp Model

Figure G.10 is the input file 741-3.CIR, which uses the uA741
macromodel as a unity-gain source follower. VIN is a squarewave which goes from 0 V to 12 V and back. The result of
slew-rate limiting can be seen in Fig. G.11 where the output
voltage, V(80), takes 23 microseconds to rise or fall 12 V.
The input squarewave has been turned into a trapezoid at the
output by the op-amp.

```
741-2.CIR  use of PSpice Model for uA741 Op-Amp
*       shows effects of driving the op-amp
*       into saturation by varying closed-loop gains
VIN  70  0  SIN(0  1  10)
RIN  70  75  RPOT 100K
.MODEL RPOT RES
.STEP LIN RES RPOT(R) .4  1  .2
RFB  80  75  1MEG
VCC  85  0  15
VEE  90  0  -15
XOP-AMP  0  75  85  90  80  UA741
*------------------------------------------------------------------------
* connections:   non-inverting input
*                 | inverting input
*                 | | positive power supply
*                 | | | negative power supply
*                 | | | | output
*                 | | | | |
.subckt uA741    1 2 3 4 5
  c1   11  12  8.661E-12
  c2    6   7  30.00E-12
  dc    5  53  dx
  de   54   5  dx
  dlp  90  91  dx
  dln  92  90  dx
  dp    4   3  dx
  egnd 99   0  poly(2) (3,0) (4,0) 0 .5 .5
  fb    7  99  poly(5) vb vc ve vlp vln 0 10.61E6 -10E6 10E6 10E6 -10E6
  ga    6   0  11 12 188.5E-6
  gcm   0   6  10 99 5.961E-9
  iee  10   4  dc 15.16E-6
  hlim 90   0  vlim 1K
  q1   11   2  13 qx
  q2   12   1  14 qx
  r2    6   9  100.0E3
  rc1   3  11  5.305E3
  rc2   3  12  5.305E3
  re1  13  10  1.836E3
  re2  14  10  1.836E3
  ree  10  99  13.19E6
  ro1   8   5  50
  ro2   7  99  100
  rp    3   4  18.16E3
  vb    9   0  dc 0
  vc    3  53  dc 1
  ve   54   4  dc 1
  vlim  7   8  dc 0
  vlp  91   0  dc 40
  vln   0  92  dc 40
.model dx D(Is=800.0E-18 Rs=1)
.model qx NPN(Is=800.0E-18 Bf=93.75)
.ends
*------------------------------------------------------------------------
.TRAN  2M  200M  0  2M
.PROBE
.END
```

Figure G.8 Input File 741-2.CIR.

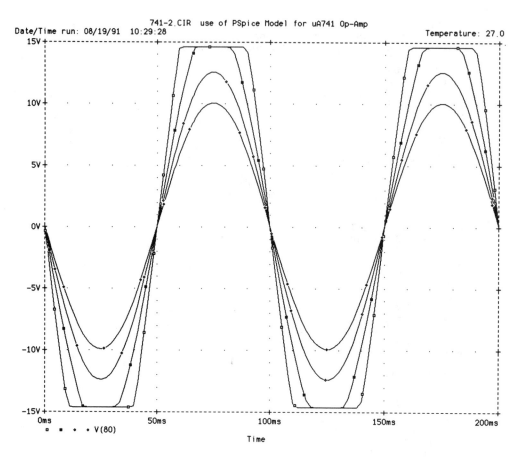

Figure G.9 Op-Amp Circuit Output Voltage with Varying Closed-Loop Gains.

G.5.4   Overshoot With PSpice Op-Amp Model

Input file 741-3.CIR, which had a 12 Vpp squarewave, was modified to be a 10 mVpp squarewave in Fig. G.12. The overshoot caused by this small-signal input to the op-amp output is apparent in Fig. G.13 in both the rise and fall, where the output overshoots by about 6%.

```
741-3.CIR  use of PSpice Model for uA741 Op-Amp
*        Response of the op-amp, as a voltage follower
*        to a large step input, graphed in the time domain
VIN   70   0   PULSE(0  12  1U  1N  1N  30U)
RL    80   0   10K
VCC   85   0   15
VEE   90   0   -15
XOP-AMP  70  80  85  90  80  UA741
*------------------------------------------------------------------------
* connections:     non-inverting input
*                  |  inverting input
*                  |  |  positive power supply
*                  |  |  |  negative power supply
*                  |  |  |  |  output
*                  |  |  |  |  |
.subckt uA741      1  2  3  4  5
  c1    11  12 8.661E-12
  c2     6   7 30.00E-12
  dc     5  53 dx
  de    54   5 dx
  dlp   90  91 dx
  dln   92  90 dx
  dp     4   3 dx
  egnd  99   0 poly(2) (3,0) (4,0) 0 .5 .5
  fb     7  99 poly(5) vb vc ve vlp vln 0 10.61E6 -10E6 10E6 10E6 -10E6
  ga     6   0 11 12 188.5E-6
  gcm    0   6 10 99 5.961E-9
  iee   10   4 dc 15.16E-6
  hlim  90   0 vlim 1K
  q1    11   2 13 qx
  q2    12   1 14 qx
  r2     6   9 100.0E3
  rc1    3  11 5.305E3
  rc2    3  12 5.305E3
  re1   13  10 1.836E3
  re2   14  10 1.836E3
  ree   10  99 13.19E6
  ro1    8   5 50
  ro2    7  99 100
  rp     3   4 18.16E3
  vb     9   0 dc 0
  vc     3  53 dc 1
  ve    54   4 dc 1
  vlim   7   8 dc 0
  vlp   91   0 dc 40
  vln    0  92 dc 40
.model dx D(Is=800.0E-18 Rs=1)
.model qx NPN(Is=800.0E-18 Bf=93.75)
.ends
*------------------------------------------------------------------------
.TRAN  .5U  60U  0  .5U
.PROBE
.END
```

Figure G.10 Input File 741-3.CIR.

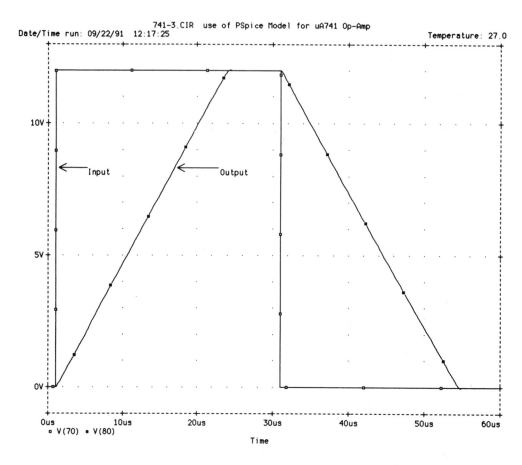

**Figure G.11 Op-Amp Circuit Input and
Output Showing Slew-Rate Limiting.**

G.6 SOURCES OF OP-AMP MACROMODELS

In recent years manufacturers of linear integrated circuits
have made available to circuit designers SPICE macromodels of
their op-amps. They are provided to assist the designer in
selecting an appropriate op-amp for a particular application,
and in fine-tuning a circuit during the early stages of the
design procedure. The following are some of the companies
that have made disks of SPICE macromodels available:

```
741-4.CIR   use of PSpice Model for uA741 Op-Amp
*        Response of the op-amp, as a voltage follower
*        to a small step input will be graphed in the time domain
VIN   70  0   PULSE(0  10m  .2U  1N  1N  2U)
RL    80  0   10K
VCC   85  0   15
VEE   90  0   -15
XOP-AMP  70  80  85  90  80  UA741
*-------------------------------------------------------------------
* connections:    non-inverting input
*                 |  inverting input
*                 |  |  positive power supply
*                 |  |  |  negative power supply
*                 |  |  |  |  output
*                 |  |  |  |  |
.subckt uA741     1  2  3  4  5
*
  c1   11  12  8.661E-12
  c2    6   7  30.00E-12
  dc    5  53  dx
  de   54   5  dx
  dlp  90  91  dx
  dln  92  90  dx
  dp    4   3  dx
  egnd 99   0  poly(2) (3,0) (4,0) 0 .5 .5
  fb    7  99  poly(5) vb vc ve vlp vln 0 10.61E6 -10E6 10E6 10E6 -10E6
  ga    6   0  11 12 188.5E-6
  gcm   0   6  10 99 5.961E-9
  iee  10   4  dc 15.16E-6
  hlim 90   0  vlim 1K
  q1   11   2  13 qx
  q2   12   1  14 qx
  r2    6   9  100.0E3
  rc1   3  11  5.305E3
  rc2   3  12  5.305E3
  re1  13  10  1.836E3
  re2  14  10  1.836E3
  ree  10  99  13.19E6
  ro1   8   5  50
  ro2   7  99  100
  rp    3   4  18.16E3
  vb    9   0  dc 0
  vc    3  53  dc 1
  ve   54   4  dc 1
  vlim  7   8  dc 0
  vlp  91   0  dc 40
  vln   0  92  dc 40
.model dx D(Is=800.0E-18 Rs=1)
.model qx NPN(Is=800.0E-18 Bf=93.75)
.ends
*-------------------------------------------------------------------
.TRAN  .01U  4U  0  .01U
.PROBE
.END
```

Figure G.12   Input File 741-4.CIR.

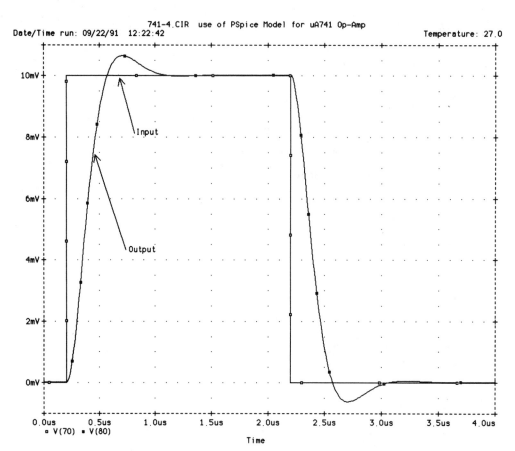

□ V(70) ■ V(80)

Time

**Figure G.13 Op-Amp Circuit Input and
Output Showing Overshoot.**

Analog Devices
70 Shawmut Road
Canton, MA   02021-3434   (617) 461-3392

Burr-Brown Corporation
P.O. Box 11400
Tucson, AZ   85734   (800) 548-6132

Linear Technology Corporation
1630 McCarthy Blvd.
Milpitas, CA   95035-7487   (408) 432-1900

Texas Instruments Incorporated
Literature Response Center
P.O. Box 809066
Dallas, TX   75380-9066   .(214) 997-3389

INDEX

305

The Standard for Circuit Simulation

FREE SOFTWARE

PSpice is a circuit simulator package used to calculate the behavior of electrical circuits. Class instructors can receive complimentary evaluation versions for *both* the IBM-PC and Macintosh by submitting a request on company or educational letterhead to:

> Product Marketing Dept.
> MicroSim Corporation
> 20 Fairbanks
> Irvine, CA 92718

Duplication of the diskettes for your students is encouraged.

PSpice is a registered trademark of MicroSim Corporation.